Ilya Gertsbakh

Reliability Theory

Springer
Berlin
Heidelberg
New York
Barcelona
Hong Kong
London
Milano
Paris
Singapore
Tokyo

Ilya Gertsbakh

Reliability Theory

With Applications to Preventive Maintenance

With 51 Figures

Professor Ilya Gertsbakh
Ben Gurion University
Department of Mathematics
P.O. Box 653
84105 Beersheva
Israel

CIP-Data applied for
Die Deutsche Bibliothek - CIP-Einheitsaufnahme
Gercbach,Il'ja B.:
Reliability theory: with applications to preventive maintenance / Ilya Gertsbakh
Berlin; Heidelberg; New York; Barcelona; Hong Kong; London; Milan; Paris; Singapore; Tokyo:
Springer, 2000
(Engineering online library)
ISBN 3-540-67275-3

ISBN 3540-67275-3 Springer-Verlag Berlin Heidelberg New York

Springer-Verlag Berlin Heidelberg New York
a member of BertelsmannSpringer Science+Business Media GmbH

Printed in Germany

Typesetting: Camera-ready copy from author
Cover-Design: MEDIO Innovative Medien Service GmbH, Berlin
Printed on acid-free paper SPIN 10763074 62/3020 5 4 3 2 1 0

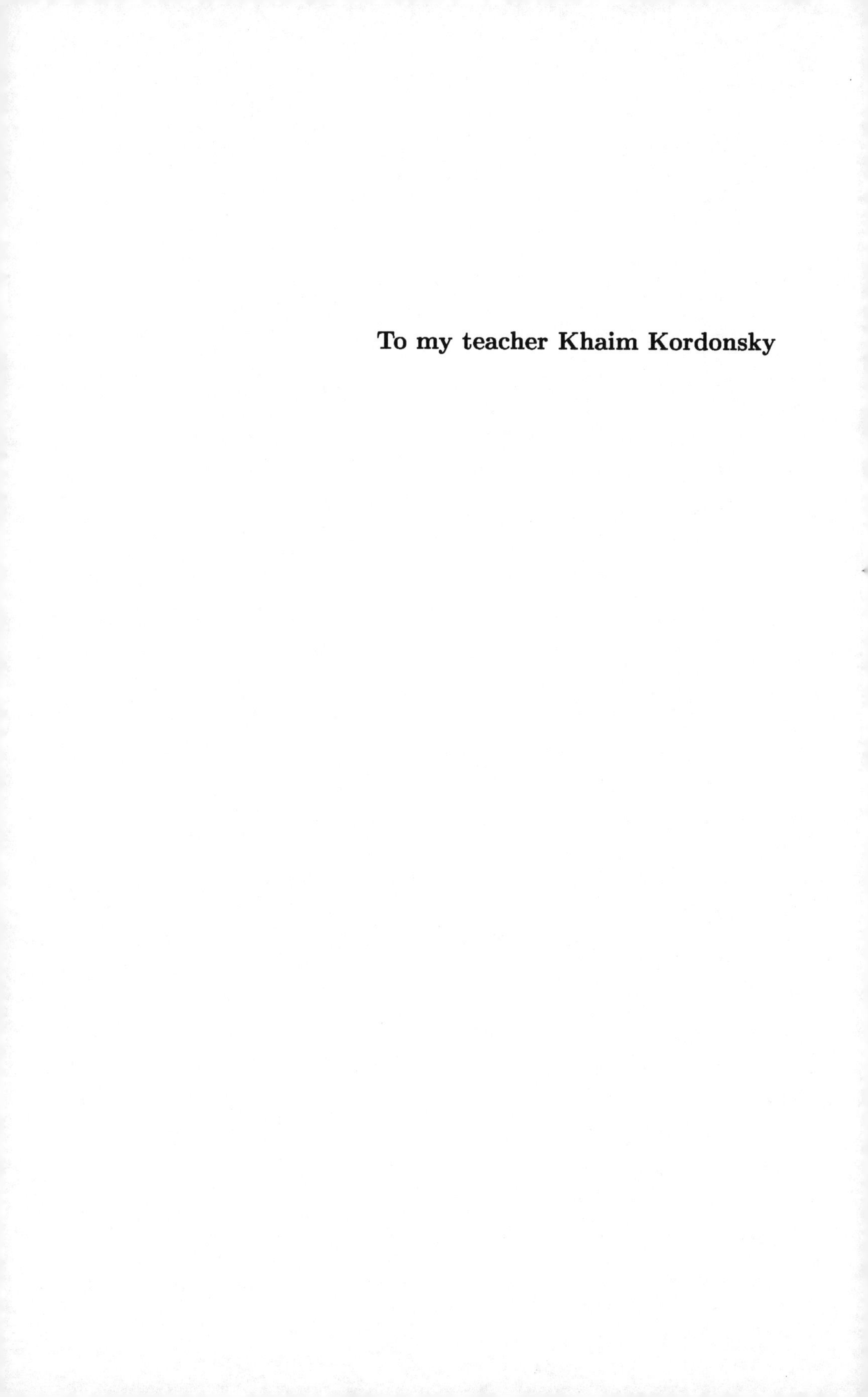

To my teacher Khaim Kordonsky

Preface

The material in this book was first presented as a one-semester course in Reliability Theory and Preventive Maintenance for M.Sc. students of the Industrial Engineering Department of Ben Gurion University in the 199 7/98 and 1998/99 academic years.

Engineering students are mainly interested in the applied part of this theory. The value of preventive maintenance theory lies in the possibility of its implementation, which crucially depends on how we handle statistical reliability data. The very nature of the object of reliability theory – system lifetime – makes it extremely difficult to collect large amounts of data. The data available are usually incomplete, e.g. heavily censored. Thus, the desire to make the course material more applicable led me to include in the course topics such as modeling system lifetime distributions (Chaps. 1, 2) and the maximum likelihood techniques for lifetime data processing (Chap. 3).

A course in the theory of statistics is a prerequisite for these lectures. Standard courses usually pay very little attention to the techniques needed for our purpose. A short summary of them is given in Chap. 3, including widely used probability plotting.

Chapter 4 describes the most useful and popular models of preventive maintenance and replacement. Some practical aspects of applying these models are addressed, such as treating uncertainty in the data, the role of data contamination and the opportunistic scheduling of maintenance activities.

Chapter 5 presents the maintenance models which are based on monitoring a "prognostic" parameter. Formal treatment of these models requires using some basic facts from Markov-type processes with rewards (costs). In recent years, there has been a growing interest in maintenance models based on monitoring the process of damage accumulation. A good example is the literature dealing with the preventive maintenance of such "nontypical" objects as bridges, concrete structures, pipelines, dams, etc. The chapter concludes by considering a general methodology for planning preventive maintenance when a system has a multidimensional state parameter. The main idea is to make the maintenance decisions depending on the value of a one-dimensional system "health index."

The material of Chap. 6 is new for a traditional course. It is based on the recent works of Kh. Kordonsky and considers the choice of the best *time scale* for age replacement. It would not be an exaggeration to say that the correct choice of the time scale is a central issue in any sphere of reliability applications.

Chapter 7 shows an example of *learning* in the process of servicing a system. Several strong assumptions were made to make the mathematics as simple as possible. It is important to demonstrate to students that the combination of prior knowledge with new data received in the process of decision making is, in fact, a universal phenomenon, which may have various useful applications.

It takes me, on the average, two weeks in the classroom (3 hours weekly) to deliver the material of one chapter. In addition, I spend some time explaining the most useful procedures of *Mathematica* needed for the numerical analysis of the theoretical models and for solving the exercises. Getting to the "real" numbers and graphs always gives students a good feeling and develops better intuition and understanding, especially if the material is saturated with statistical notions. The course concludes with detailed solutions of the exercises, including a numerical investigation by means of *Mathematica*.

Ilya Gertsbakh
Beersheva, January 2000

Contents

Chapter 1

System Reliability as a Function of Component Reliability

The whole is simpler than the sum of its parts
Willard Gibbs

1.1 The System and Its Components

In reliability theory, as in any theory, we think and operate in terms of *models*. In this chapter we investigate a model of a *system*, which consists of *elements* or *components*. Our purpose is to develop a formal instrument to enable us to receive information about a system's reliability from information about the reliability of its components. The exposition in this section does not involve probabilistic notions.

A system is a set of components (elements). Only *binary* components will be considered, i.e. components having only two states: operational (up) and failed (down). The state of component i, $i = 1, \ldots, n$, will be described by a binary variable x_i: $x_i = 1$ if the component is up; $x_i = 0$ if the component is down.

It will be assumed that the whole system can only be in one of two states: up or down. The dependence of a system's state on the state of its components will be determined by means of the so-called *structure* function $\phi(\mathbf{x})$, where $\mathbf{x} = (x_1, x_2, \ldots, x_n)$: $\phi(\mathbf{x}) = 1$ if the system is up; $\phi(\mathbf{x}) = 0$ if the system is down.

We also use the notation $\mathbf{x} < \mathbf{y}$. This means that the components of $\mathbf{x}$ are less then or equal to the components of $\mathbf{y}$, i.e. $x_i \le y_i$, but for at least one component j, $x_j < y_j$.

Example 1.1.1: Series system (Fig. 1.1a)
This system is up if and only if all its components are up. Formally,

$$\phi(\mathbf{x}) = \prod_{i=1}^{n} x_i = \min_{1 \le i \le n} x_i. \tag{1.1.1}$$

Example 1.1.2: Parallel system (Fig. 1.1b)
The system is up if and only if at least one of its components is up. Formally,

$$\phi(\mathbf{x}) = 1 - \prod_{i=1}^{n}(1 - x_i) = \max_{1 \le i \le n} x_i. \ \# \tag{1.1.2}$$

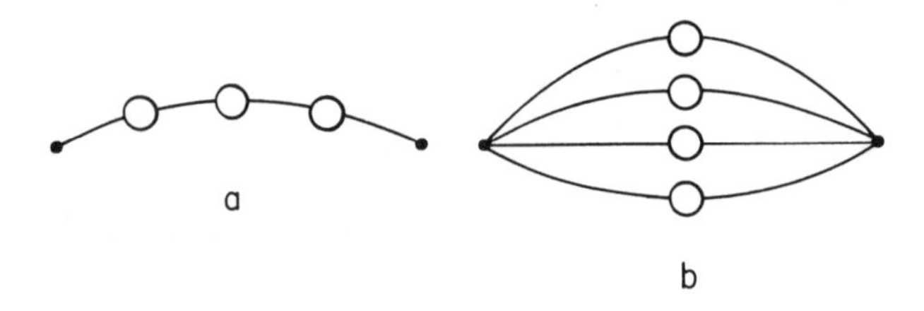

Figure 1.1. Representation of series (a) and parallel system (b)

Example 1.1.3: k-out-of-n system
This system is up if and only if at least k out of its n components are operating. Formally,

$$\phi(\mathbf{x}) = 1, \ \text{if} \ \sum_{i=1}^{k} x_i \ge k, \tag{1.1.3}$$

and $\phi(\mathbf{x}) = 0$ otherwise
Example 1.1.4: Cable TV transmitter (Fig. 1.2)
The system is designed to transmit from the central station S to three local stations S_1, S_2, S_3. The stations are connected by cables numbered $1, 2, 3, 4, 5$, which are the system components. The system is operational (up) if all substations are connected directly or through another substation to the central station.

One can check that

$$\begin{aligned}\phi(\mathbf{x}) &= 1 - (1 - x_2x_3x_5)(1 - x_2x_4x_5) \\ &\times (1 - x_2x_3x_4)(1 - x_1x_3x_4)(1 - x_1x_3x_5)(1 - x_1x_2x_5)(1 - x_1x_2x_4).\end{aligned}$$

We explain later how to derive this formula.

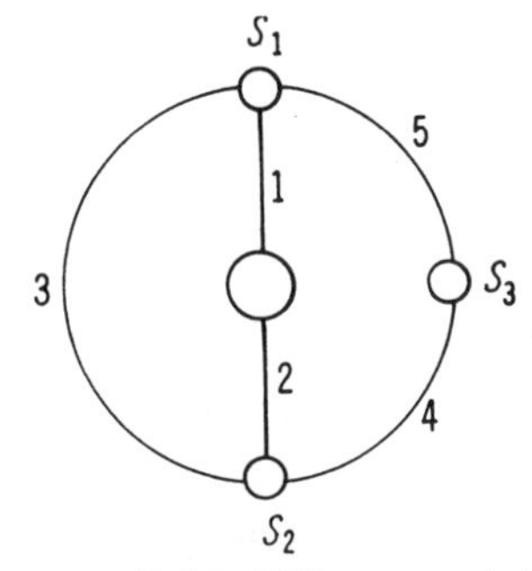

Figure 1.2. Cable TV transmission system

Example 1.5: Series connection of parallel systems (Fig. 1.3)
For this system, $\phi(\mathbf{x}) = [1-(1-x_1)(1-x_2][1-(1-x_3)(1-x_4)]$.

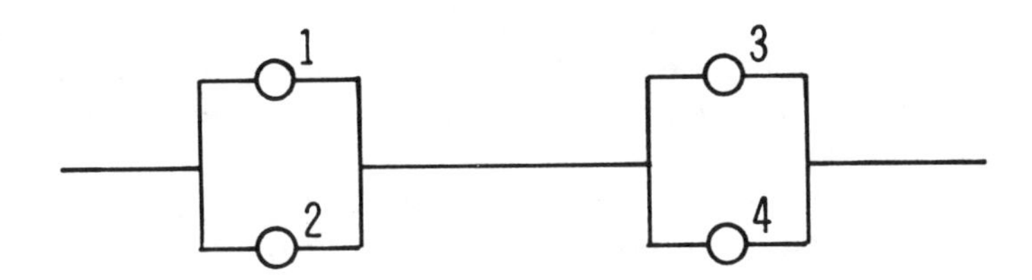

Fig. 1.3. Series connection of parallel systems

Example 1.1.6: Parallel connection of series systems (Fig. 1.4)
Check that for this system $\phi(\mathbf{x}) = 1-(1-x_1x_2)(1-x_3x_4x_5)$.

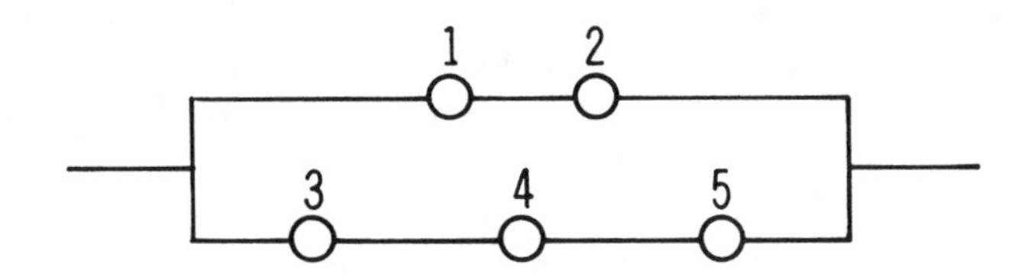

Figure 1.4. Parallel connection of series systems

It is important to have a systematic way of constructing a formula for the structure function $\phi(\cdot)$. This will be done by using the notions of ***minimal paths*** and *minimal cuts.* Before doing this let us impose some natural demands on $\phi(\mathbf{x})$.

Definition 1.1.1: *Monotone system*
A system with structure function $\phi(\cdot)$ is called ***monotone*** if it has the following properties:

(i) $\phi(0,0,\ldots,0)=0, \quad \phi(1,1,\ldots,1)=1$;

(ii) $\mathbf{x}<\mathbf{y} \Rightarrow \phi(\mathbf{x}) \leq \phi(\mathbf{y})$.

In words: the system is down if all its elements are down; it is up if all its elements are up; and the state of the system cannot become worse if any of its elements changes its state from down to up.

Definition 1.1.2: *Cut vector, cut set, path vector, path set*
A state vector $\mathbf{x}$ is called a *cut vector* if $\phi(\mathbf{x})=0$. The set $C(\mathbf{x})=\{i: x_i=0\}$ is then called a *cut set.* If, in addition, for any $\mathbf{y}>\mathbf{x}, \ \phi(\mathbf{y})=\mathbf{1}$, then the corresponding cut set is called ***minimal cut set*** or simply minimal cut.

A state vector $\mathbf{x}$ is then called a ***path vector*** if $\phi(\mathbf{x})=1$. The set $A(\mathbf{x})=\{i: x_i=1\}$ is then called a ***path set.*** If, in addition, for any $\mathbf{y}, \ \mathbf{y}<\mathbf{x}, \phi(\mathbf{y})=0$, then the corresponding path set is called ***minimal path set,*** or minimal path.

A minimal cut set is a minimal set of components whose failure causes the failure of the whole system.

If all elements of the path set are "up" then the system is up. A minimal path is a minimal set of elements whose functioning (i.e. being up) ensures that the system is up. The minimal path set cannot be reduced, as it has no redundant elements.

Examples 1.1.5, 1.1.6 continued
For Example 1.1.5, $\mathbf{x}_1=(1,1,1,0)$ is a path vector. The corresponding path set is $\{1,2,3\}$. It is *not,* however, a minimal path set because if element 2 is turned down the system will still be up. $\{1,3\}$ is the minimal path set. There are three other minimal path sets. Find them!

For Example 1.1.6, there are two minimal path sets, $\{1,2\}$ and $\{3,4,5\}$.

The system in Fig. 1.3 has two minimal cuts: $\{1,2\}$ and $\{3,4\}$.

The set $\{1,2,3\}$ is also a cut set but not a minimal one. The system in Fig. 1.4 has six minimal cuts of the form $\{i,j\}$, where $i=1,2$ and $j=3,4,5$.

Theorem 1.1.1: *Structure function representation*
Let $P_1, P_2, \ldots, P_s$ be the minimal path sets of the system. Then

$$\phi(\mathbf{x})=1-\prod_{j=1}^{s}\Big(1-\prod_{i\in P_j} x_i\Big) . \tag{1.1.4}$$

Let $C_1, C_2, \ldots, C_k$ be the minimal cut sets of the system. Then

$$\phi(\mathbf{x}) = \prod_{j=1}^{k} \left(1 - \prod_{i \in C_j} (1 - x_i)\right) . \tag{1.1.5}$$

Proof
Assume that there is at least one minimal path set, all elements of which are up, say P_1. Then $\prod_{i \in P_1} x_i = 1$ and this leads to $\phi(\mathbf{x}) = 1$. Suppose now that the system is up. Then there must be one minimal path set having all of its elements in the up state. Thus the right-hand side of (1.1.4) is 1. Therefore, $\phi(\mathbf{x}) = 1$ if and only if there is one minimal path set having all its elements in the up state. This proves (1.1.4).

We omit the proof of (1.1.5), which is similar.

It follows from Theorem 1.1.1 that any monotone system can be represented in two equivalent ways: as a series connection of parallel subsystems each being a minimal cut set, or as a parallel connection of series subsystems each being a minimal path set. Therefore, there are two ways to represent structure functions. After corresponding simplifications, these become identical, as the following example shows.

Example 1.1.5 continued
The structure function given above for the system in Fig. 1.3 is based on minimal cuts $\{1,2\}$ and $\{3,4\}$. The system also has four minimal paths: $\{1,3\}, \{1,4\}$, $\{2,3\}, \{2,4\}$. Thus, $\phi(\mathbf{x}) = 1 - (1 - x_1x_3)(1 - x_1x_4)(1 - x_2x_3)(1 - x_2x_4)$.

The structure function based on minimal cuts was presented in Example 1.1.5. Both formulas produce identical results. To verify this, it is necessary to simplify both expressions. Note that for binary variables, $x_i^k = x_i$ for any integer k.

More information on monotone systems and their structure function can be found in the literature, e.g. in Barlow and Proschan (1975), Chap. 1, and Gertsbakh (1989), Chap. 1.

1.2 Independent Components: System Reliability and Stationary Availability

Contrary to Sect. 1.1, let us now assume that the state of component i is described by a binary *random* variable X_i, defined by

$$P(X_i = 1) = p_i, \quad P(X_i = 0) = 1 - p_i, \tag{1.2.1}$$

where 1 and 0 correspond to the operational (up) and failure (down) state, respectively.

It will be assumed that all components are mutually independent. This implies a considerable formal simplification: for independent components, the joint distribution of $X_1, X_2, \ldots, X_n$ is completely determined by component reliabilities $p_1, p_2, \ldots, p_n$.

Denote by $\mathbf{X} = (X_1, X_2, \ldots, X_n)$ the system state vector. This is now a random vector. Correspondingly, the system structure function $\phi(\mathbf{X}) = \phi(X_1, \ldots, X_n)$ becomes a binary random variable: $\phi(\mathbf{X}) = 1$ corresponds to the system up state and $\phi(\mathbf{X}) = 0$ corresponds to the system down state.

Definition 1.2.1: *System reliability*
System reliability r_0 is the probability that the system structure function equals 1:

$$r_0 = P(\phi(\mathbf{X}) = 1) . \tag{1.2.2}$$

Since $\phi(\cdot)$ is a binary random variable, the last formula can be written as

$$r_0 = E[\phi(\mathbf{X})] . \tag{1.2.3}$$

Expression (1.2.3) is very useful since the operation of taking expectation $E[\cdot]$ is a very powerful tool for reliability calculations. The following examples show how to compute system reliability via its structure function.

Example 1.2.1: Reliability of a series system
$\phi(\mathbf{X}) = \prod_{i=1}^{n} X_i$, and therefore

$$r_0 = E[\phi(\mathbf{X})] = \prod_{i=1}^{n} p_i. \tag{1.2.4}$$

Example 1.2.2: Parallel system
Here $\phi(\mathbf{X}) = 1 - \prod_{i=1}^{n}(1 - X_i)$. Thus

$$r_0 = E[\phi(\mathbf{X})] = 1 - \prod_{i=1}^{n}(1 - p_i) . \tag{1.2.5}$$

Example 1.2.3: Series connection of parallel systems (Example 1.1.5)
From the expression for $\phi(\mathbf{X})$ it follows immediately that

$$r_0 = E[\phi(\mathbf{X})] = [1 - (1 - p_1)(1 - p_2)][1 - (1 - p_3)(1 - p_4)] . \tag{1.2.6}$$

Example 1.2.4: 2-out-of-4 system with identical elements
For this system,

$$r_0 = E[\phi(\mathbf{X})] = P(\sum_{i=1}^{4} X_i \geq 2) = \sum_{m=2}^{4} \binom{n}{m} p^m (1 - p)^{4-m} . \tag{1.2.7}$$

Let $(\alpha_i; \mathbf{p})$ denote the vector $\mathbf{p}$ with its ith component replaced by α_i. So $(1_i; \mathbf{p}) = (p_1, \ldots, p_{i-1}, 1, p_{i+1}, \ldots, p_n)$.

Theorem 1.2.1: *Pivotal decomposition*
Let $r_0 = r(\mathbf{p}) = r(p_1, \ldots, p_n)$ be the reliability of a monotone system with independent components. Then

$$r(\mathbf{p}) = p_i r(1_i; \mathbf{p}) + (1 - p_i) r(0_i; \mathbf{p}). \tag{1.2.8}$$

Proof
By definition, $r_0 = E[\phi(\mathbf{X})] = P[\phi(\mathbf{X}) = 1]$. By the formula of total probability,
$r_0 = p_i P(\phi(\mathbf{X}) = 1 | X_i = 1) + (1 - p_i) P(\phi(\mathbf{X}) = 1 | X_i = 0)$, or
$r_0 = p_i P(\phi(1_i; \mathbf{X}) = 1 | X_i = 1) + (1 - p_i) P(\phi(0_i; \mathbf{X}) = 1 | X_i = 0)$.
Since $X_1, \ldots, X_n$ are independent, the last equality takes the form:

$$r_0 = p_i P(\phi(1_i; \mathbf{X}) = 1) + (1 - p_i) P(\phi(0_i; \mathbf{X}) = 1). \tag{1.2.9}$$

Expression (1.2.9) can be rewritten as

$$r(\mathbf{p}) = p_i r(1_i; \mathbf{p}) + (1 - p_i) r(0_i; \mathbf{p}) \,. \tag{1.2.10}$$

Physically, $r(1_i; \mathbf{p})$ is the reliability of a system in which the ith component is replaced by an absolutely reliable one; similarly, $r(0_i; \mathbf{p})$ is the reliability of the system in which the ith component has failed. Pivoting (1.2.10) is applied until replacement of components by absolutely reliable ones and/or by failed ones produces a structure for which the reliability is easily computable, such as the series-parallel system. Let us demonstrate how to use the pivotal formula to compute the reliability of the bridge structure shown in Fig. 1.5.

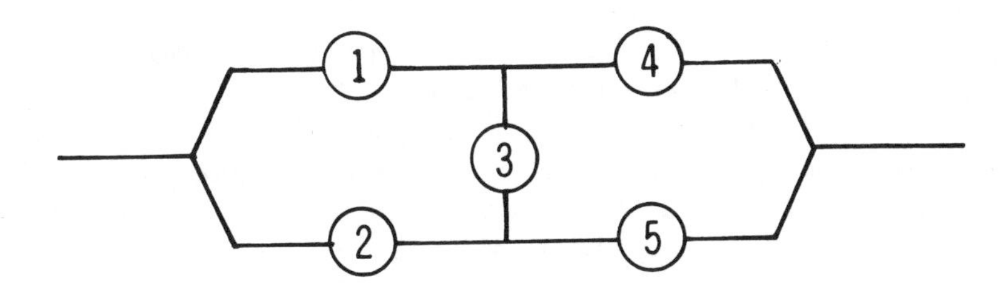

Figure 1.5. Bridge structure

Example 1.2.5: Reliability of a bridge structure
The best choice is to pivot around element 3. Suppose that component 3 is up. Then the bridge becomes a series connection of two parallel subsystems consisting of elements 1, 2 and 4, 5, respectively. Its reliability (cf. (1.2.6)) is
$r(1_3; \mathbf{p}) = [1 - (1 - p_1)(1 - p_2)][1 - (1 - p_4)(1 - p_5)]$.

If 3 is down, then the bridge becomes a parallel connection of two series systems: one with components 1, 4 and the second with components 2, 5. Its reliability is $r(0_3;\mathbf{p}) = 1-(1-p_1p_4)(1-p_2p_5)$. Therefore, the system reliability is $r_0 = p_3 \cdot r(1_3;\mathbf{p}) + (1-p_3)r(0_3;\mathbf{p})$.

It is easy to finish the calculations. The final result is

$$\begin{aligned} r_0 = \quad & E[\phi(\mathbf{X})] = p_1p_3p_5 + p_2p_3p_4 + p_2p_5 + p_1p_4 - p_1p_2p_3p_5 \\ & -p_1p_2p_4p_5 - p_1p_3p_4p_5 - p_1p_2p_3p_4 - p_2p_3p_4p_5 + \\ & 2p_1p_2p_3p_4p_5. \end{aligned} \quad (1.2.11)$$

It is instructive to obtain the same result by using, for example, minimal paths. The bridge has four minimal path sets: $\{1,3,5\}$, $\{2,3,4\}$, $\{1,4\}$, and $\{2,5\}$. Thus the random structure function is

$$\phi(\mathbf{X}) = 1-(1-X_1X_3X_5)(1-X_2X_3X_4)(1-X_2X_5)(1-X_1X_4). \quad (1.2.12)$$

Expand the terms in parentheses, simplify the expression using the fact that $X_i^k = X_i$ and take the expectation.

More on the use of the pivotal decomposition can be found in Gertsbakh (1989) and Barlow (1998) and the references there.

System reliability r_0 as defined above is of a purely "static" nature: system operation *time* is not present at all. One can imagine a coin being flipped for element i showing "up" and "down" with probabilities p_i and $1-p_i$, respectively. Then, for this static experiment, r_0 is the probability that the system will be in the "up" state.

Another interpretation might be the following. Assume that we have a certain instant t^* on the time axis and p_i is the probability that component i is up at t^*. Then by (1.2.2) r_0 represents the probability that the system is up at t^*. We will show in the next section how this fact leads to an expression for the system lifetime distribution function.

There is another interpretation of the quantity r_0 which involves time and is related to the so-called ***system availability***. First, suppose that each component has alternating up and down periods on the time axis. The whole system also has alternating up and down periods on the time axis. Figure 1.6 illustrates this situation for a series system of two components. System up periods are those for which both components are in the up state.

Consider component k and denote by $\xi_j^{(k)}$ and $\eta_j^{(k)}$, $j = 1,2,3,\ldots$, the sequential up and down periods for this component. Suppose that $\xi_j^{(k)} + \eta_j^{(k)}, j = 1,2,\ldots$, are i.i.d. random variables. The following quantity is called the ***stationary availability*** of component k:

$$A_v^{(k)} = \frac{\mu^{(k)}}{\mu^{(k)} + \nu^{(k)}}, \quad (1.2.13)$$

where $\mu^{(k)} = E[\xi_j^{(k)}]$, $\nu^{(k)} = E[\eta_j^{(k)}]$.

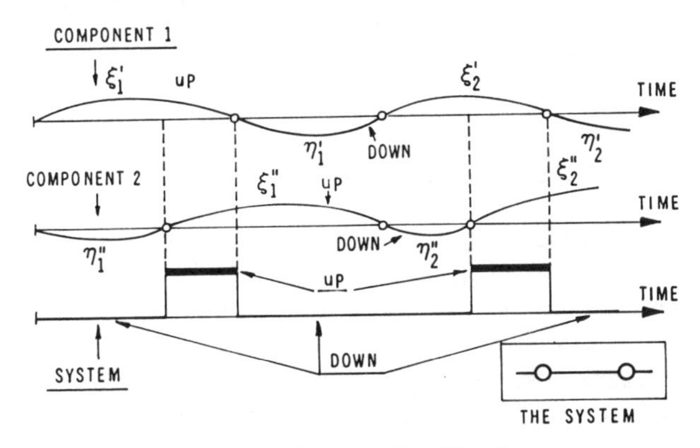

Figure 1.6. The system is up if and only if both components are up

$A_v^{(k)}$ has the following probabilistic interpretation. Let $P^{(k)}(t)$ be the probability that component k is up at some time instant t. Then

$$A_v^{(k)} = \lim_{t \to \infty} P^{(k)}(t). \tag{1.2.14}$$

One can say that the stationary availability is the probability that the component is up at some remote instant of time (formally, as $t \to \infty$). Another interpretation is the following. Denote by $V^{(k)}(T)$ the total amount of up time for component k in $[0, T]$. Then

$$A_v^{(k)} = \lim_{T \to \infty} E[V^{(k)}(T)]/T. \tag{1.2.15}$$

Now we are able to claim the following. Suppose that we have a formula expressing the "static" up probability r_0 of the system as a function of component up probabilities p_i, i.e. $r_0 = \psi(p_1, p_2, \ldots, p_n)$. (Recall that the components are assumed to be independent.) Suppose that $p_k = A_v^{(k)}$, $k = 1, 2, \ldots, n$, i.e. p_k equals the stationary availability of the kth component. Then system stationary availability A_v is equal to

$$A_v = \psi(A_v^{(1)}, \ldots, A_v^{(n)}). \tag{1.2.16}$$

System stationary availability can be interpreted as the limiting probability that the system will be in the up state at a remote moment in time. Alternatively, let $V(T)$ be the total amount of up time for the system on $[0, T]$. Then $A_v = \lim_{T \to \infty} E[V(T)]/T$.

The proof of (1.2.15) can be found in, for example, Barlow and Proschan (1975), Chap. 7.

For a series system of independently operating and repaired components, A_v has the form

$$A_v = \prod_{k=1}^{n} \frac{\mu^{(k)}}{\mu^{(k)} + \nu^{(k)}}\,. \tag{1.2.17}$$

This important formula is widely used in reliability practice.

System availability will be discussed in Chap. 4 in connection with alternating renewal processes.

1.3 Lifetime Distribution of a System without Component Renewal

The following properties of system components will be assumed for the exposition in this section:

(i) Component i has lifetime τ_i with known cumulative distribution function (c.d.f.) $F_i(t) = P(\tau_i \le t)$, $i = 1, \ldots, n$.

(ii) τ_i are independent random variables (r.v.'s), $i = 1, \ldots, n$.

(iii) At $t = 0$, all components are up. A component which fails is *not* renewed or repaired. A components which fails remains in the failure (down) state "for ever."

To define system lifetime, we need some extra notation. Let a binary 0/1 r.v. $X_i(t)$ be defined as follows: $X_i(t) = 1$ if and only if $\tau_i > t$. If $\tau_i \le t$ then $X_i(t) = 0$. In other words, $X_i(t)$ equals one as long as component i is up, and becomes zero when this component goes down.

Let $\mathbf{X}(t) = (X_1(t), \ldots, X_n(t))$ be the component state vector at time t.

Definition 1.3.1: *System lifetime*
System lifetime τ is the time until the system first enters the down (failure) state:

$$\tau = \inf[t : \phi(\mathbf{X}(t)) = 0] . \tag{1.3.1}$$

We define system reliability $R(t)$ as the probability that τ exceeds t: $R(t) = P(\tau > t)$. Often $R(t)$ is called the survival probability or survival function.

Let $r_0 = E[\phi(X_1, \ldots, X_n)] = \psi(p_1, \ldots, p_n)$ be the "static" reliability of the system; see Definition 1.2.1. The following theorem tells us how to find $R(t)$ by means of the function $\psi(p_1, \ldots, p_n)$.

Denote by $R_i(t)$ the reliability of component i, i.e. $R_i(t) = P(\tau_i > t)$.

Theorem 1.3.1: *System reliability*

$$R(t) = \psi(R_1(t), R_2(t), \ldots, R_n(t)), \tag{1.3.2}$$

where $R_i(t) = 1 - F_i(t)$.

Proof
It follows from the definition of $R_i(t)$ that $R(t)$ is the probability that the system is up at time t. A monotone system consisting of nonrenewable components starts operating at $t = 0$ in the "up" state, eventually entering the "down"

state and remaining in it for ever. Thus, if the system is "up" at the instant t, it was "up" during the whole time interval $[0, t]$, or $R(t) = P(\tau > t)$.

Theorem 1.3.1 says that the system reliability function is obtained by replacing in the structure function the component reliabilities p_i by the corresponding reliability functions $R_i(t)$.

Example 1.3.1: Minimum-maximum calculus
For a series system, $r_0 = \prod_{i=1}^n p_i$. Then by Theorem 1.3.1,

$$R(t) = \prod_{i=1}^{n} (1 - F_i(t)) \; . \tag{1.3.3}$$

Let $F(t)$ be the c.d.f. of system lifetime τ. Then

$$P(\tau \le t) = F(t) = 1 - R(t) = 1 - \prod_{i=1}^{n} (1 - F_i(t)). \tag{1.3.4}$$

Let us recall how the formula for the c.d.f. of the minimum τ of n independent r.v.s $\tau_1, \ldots, \tau_n$ is derived. The events $\{\tau > t\}$ and $\{\tau_1 > t, \ldots, \tau_n > t\}$ are equivalent. Thus $P(\tau > t) = \prod_{i=1}^n P(\tau_i > t) = \prod_{i=1}^n R_i(t)$.

Therefore, the lifetime of a series system of independent components coincides with the minimum lifetime of its components.

For a parallel system, $r_0 = 1 - \prod_{i=1}^n (1 - p_i)$, and therefore

$$R(t) = 1 - \prod_{i=1}^{n} F_i(t) \; . \tag{1.3.5}$$

Similarly to the series system, it is easy to show that $1 - R(t) = \prod_{i=1}^n F_i(t)$ is the formula for the c.d.f. of the maximum of n independent r.v.'s $\tau = \max(\tau_1, \tau_2, \ldots, \tau_n)$. In other words, the lifetime of a parallel system coincides with the maximum lifetime of its components.

It is now obvious that the lifetime of a series-parallel system can be expressed via the component lifetimes using minimum and maximum operations. Let us consider, for example, the system shown in Fig. 1.7.[1]

Let $\tau_i \sim F_i(t)$. Clearly, the system lifetime is

$$\tau = \max(\tau_5, \min(\tau_1, \tau_4, \max(\tau_2, \tau_3))). \tag{1.3.6}$$

Obviously,

$$\tau^* = \max(\tau_2, \tau_3) \sim F^*(t) = F_2(t) F_3(t). \tag{1.3.7}$$

Similarly,

$$\tau^{**} = \min((\tau_1, \tau_4), \tau^*) = 1 - R_1(t) R_4(t) R^*(t) \; . \tag{1.3.8}$$

[1] Borrowed from Gertsbakh (1989), p. 15, with the permission of Marcel Dekker, Inc.

Now, $\tau = \max(\tau_5, \tau^{**})$, and by (1.3.7),

$$F(t) = F_5(t)(1 - R_1(t)R_4(t)(1 - F_2(t)F_3(t))). \tag{1.3.9}$$

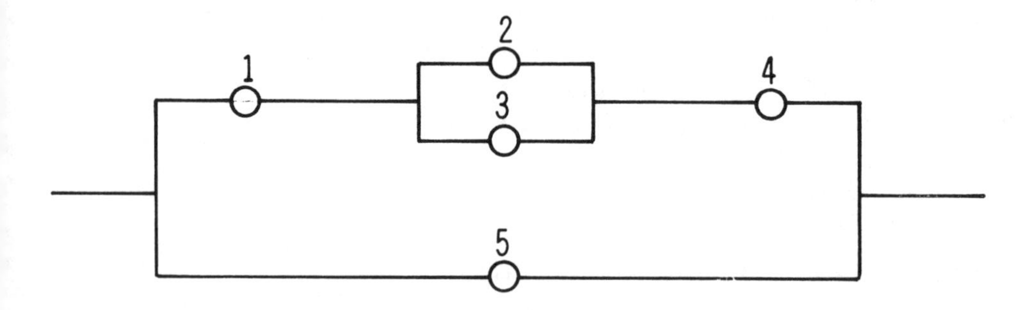

Figure 1.7. A series-parallel system

We have seen that for a series-parallel system it is quite easy to derive an expression for system lifetime distribution. It turns out that the reliability function of a monotone system with independent components can be ***approximated*** from above and from below by a reliability function of a specially chosen series-parallel system. This is summarized in the following statement; see, for example, Barlow and Proschan (1975), Chap. 2.

Theorem 1.3.2
Let $A_1, A_2, \ldots, A_l$ be the minimal path sets, and let $C_1, C_2, \ldots, C_m$ be the minimal cut sets of a monotone system. Denote the reliability of ith component by p_i, and the system reliability by r_0. Then

$$\prod_{k=1}^{m} \Big(1 - \prod_{j \in C_k} (1 - p_j)\Big) \le r_0 \le 1 - \prod_{r=1}^{l} \Big(1 - \prod_{j \in A_r} p_j\Big). \tag{1.3.10}$$

This theorem says that r_0 is bounded from above by the reliability of a fictitious system (with ***independent*** components) which is a parallel connection of series subsystems each being the minimal path set of the original system. Similarly, the lower bound is the reliability of a system obtained by a series connection of parallel subsystems each being the minimal cut set of the original system. For example, the "lower system" for a bridge is shown in Fig. 1.8a. Similarly, Fig. 1.8b shows the "upper system". Note that the elements i and i^* are assumed to be independent.

Denote the lower bound in (1.3.10) by $LB(p_1, \ldots, p_n)$ and the upper bound by $UB(p_1, \ldots, p_n)$. Then, using the arguments of Theorem 1.3.1, we can establish the following:

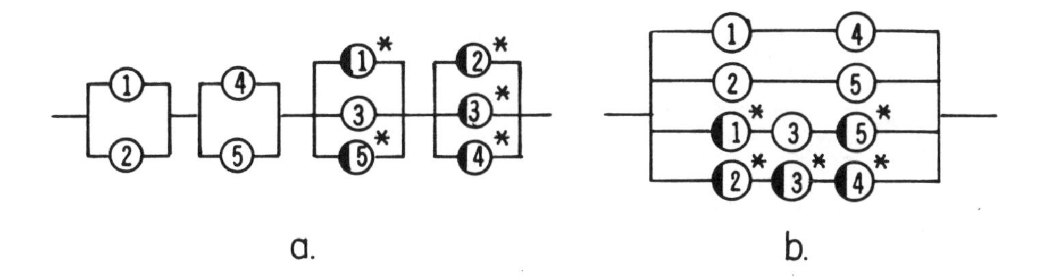

Figure 1.8. The "lower" (a) and the "upper" (b) system for the bridge

Corollary 1.3.1

$$LB(R_1(t), \ldots, R_n(t)) \leq R(t) \leq UB(R_1(t), \ldots, R_n(t)) \,. \tag{1.3.11}$$

In other words, the system reliability function $R(t)$ is bounded from above and from below by the reliability function of the "lower" and "upper" system, respectively.

In conclusion, let us present a very useful formula for computing the mean value of system lifetime. Suppose that τ is a ***nonnegative*** random variable, i.e. $P(\tau \geq 0) = 1$. Let $F(t)$ be the c.d.f. of τ, i.e. $P(\tau \leq t) = F(t)$. Then

$$E[\tau] = \int_0^\infty (1 - F(t))dt. \tag{1.3.12}$$

To prove this formula, integrate $\int_0^\infty (1 - F(t))dt$ by parts and use the fact that if $E[\tau]$ exists, then $\lim_{t\to\infty}[t(1 - F(t))] = 0$.

Example 1.3.2
Let $\tau \sim F(t)$ and let T be a constant. Define $X = \min[\tau, T]$. Find $E[X]$.

The physical meaning of the r.v. X is the following. Suppose that τ is the system lifetime. We stop the system either at its failure or at "age" T, whichever occurs first. This situation is typical for the so-called *age replacement* which we will be considering later; see Chap. 4. Then X is the random time span during which the system has operated before it was stopped.

Note that T can be viewed as a discrete random variable whose all probability mass is concentrated in the point $t = T$. T has the following c.d.f.: $F_T(t) =$

$P(t \le T) = 0$ for $t < T$, and $F_T(t) = 1$ for $t \ge T$. Moreover, T is independent of τ. Then by (1.3.3)

$$F_X(t) = (1 - F(t)) \cdot (1 - F_T(t))dt. \quad (1.3.13)$$

Since $F_X(t) = 0$ for $t \ge T$,

$$E[X] = \int_0^T (1 - F(t))dt \, . \quad (1.3.14)$$

1.4 Exercises

1. Consider the following system:

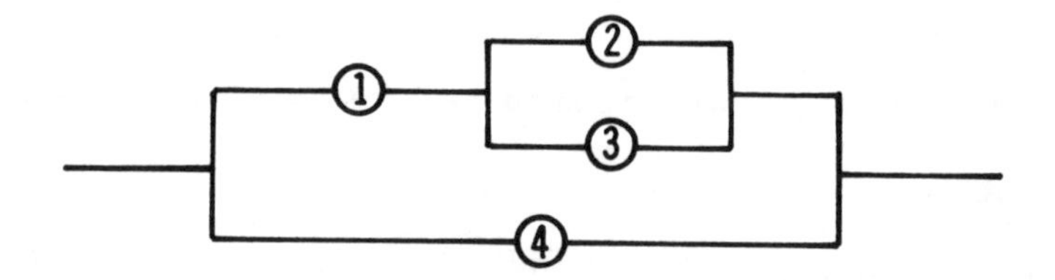

All elements are independent. Denote by p_i the reliability of element i, $i = 1, 2, 3, 4$.

a. Find $\phi(\mathbf{x})$, the system structure function.

b. Find $r_0 = E[\phi(\mathbf{X})]$, the system reliability.

c. Find the Barlow–Proschan upper and lower bounds on r_0.

d. Suppose that the lifetime τ_i of element i is exponential, i.e. $P(\tau_i \le t) = 1 - e^{-\lambda_i t}$. Find the system lifetime distribution function.

e. Suppose that element i is up, on average, during time $\mu_i = 10 + i$, and is down, on average, during time $\nu_i = 2 + 0.5i$. Periods of up and down times alternate for each element. Find the system stationary availability.

2. Let system reliability be $r_0 = \psi(p_1, p_2, \dots, p_n)$.

The so-called Birnbaum importance index of element i is defined as $R(i) = \partial\psi/\partial p_i$.

a. Find $R(3)$ for the system of Exercise 1. Find the element with greatest importance for $p_1 = 0.9, p_2 = 0.8, p_3 = 0.5 = p_4$.

b. Use the pivotal decomposition formula (1.2.10) to prove that $R(i) = r(1_i; \mathbf{p}) - r(0_i; \mathbf{p})$.

Hint: Differentiate $r(\mathbf{p})$ with respect to p_i.

3. Let $(1_i, \mathbf{x})$ be the vector $\mathbf{x} = (x_1, \dots, 1, \dots, x_n)$, i.e. the ith coordinate is set to 1. Similarly, $(0_i, \mathbf{x})$ is the vector $\mathbf{x}$ with ith coordinate equal to 0. Prove that $\phi\mathbf{x})$ is a linear combination of $\phi(1_i, \mathbf{x})$ and $\phi(0_i, \mathbf{x})$ with coefficients x_i and $1 - x_i$, respectively.

4. n identical elements, each with exponentially distributed lifetimes, are connected in parallel. The mean lifetime of a single element equals $1/\lambda = 1$. What value of n would make the mean lifetime of the whole system equal to at least 2?

5. Let n independent elements be connected in series. Suppose that the lifetime of component I is exponential with parameter λ_i, $i = 1, 2, \ldots, n$. Find the lifetime distribution function for the whole system.

6. Consider the bridge structure of Sect. 1.2. Assume that all elements have reliability $p = 0.9,\ 0.95,\ 0.99$. Compute the exact reliability and the Barlow–Proschan upper and lower bounds.

7. Consider a parallel and a series system with component reliabilities $p_1 \leq p_2 \leq \ldots \leq p_n$. What are the most important components for these systems, according to Birnbaum's measure?

8. Compute bridge reliability using the pivotal decomposition formula.
Hint: Pivot around component 3 in Fig. 1.5.

9. A radar system consists of three identical, independently operating stations. The system is considered to be "up" if and only if at least two stations are operational. Assume that station's lifetime $G(t) = 1 - \exp[-\lambda t]$. After the system enters the down state, it is repaired and brought to the "brand new" (up) state during time $t_{rep} = 0.1/\lambda$.

a. Find the mean duration of the "up" time.

b. Find the system stationary availability.

10. A system consists of two units, A and B. If A is up and B is down, or vice versa, the system is in the "up" state. Because of the power supply system deficiency, simultaneous operation of A and B is impossible, since there is not enough power for both units. Is this system monotone?

Chapter 2

Parametric Lifetime Distributions

Everything should be made as simple as possible, but not simpler.
Albert Einstein

2.1 Poisson Process – Exponential and Gamma Distributions

Suppose that we observe certain *events* that appear at random time instants. For example, the events might be arrivals of buses at a bus terminal, or telephone calls addressed to a certain person, or failures of a piece of equipment. To be specific, we shall refer to some particular sort of event, say telephone calls.

Denote by $N(t)$ the total number of calls which take place during the interval $(0, t]$. For each fixed t, $N(t)$ is a random variable, and a family of random variables $\{N(t),\ 0 < t < \infty\}$ is termed a *counting random process.*

In this section we restrict our attention to processes with *stationary* increments: the number of calls in the interval $(t_1+s, t_2+s]$ has the same distribution as the number of calls in any other interval of length $(t_2 - t_1)$.

One very important process of this type is the so-called Poisson process. Denote by Δ_i the interval $(t, t+\Delta_i]$ and let $A_i(k_i)$ be the event "the number of calls during Δ_i equals k_i." Denote by $N(h)$ the number of events in the interval $(t, t+h]$.

Definition 2.1.1
$\{N(t),\ t > 0\}$, is said to be a Poisson process with rate λ, if:

(i) for any set of nonoverlapping intervals $\Delta_1, \ldots, \Delta_m$, and for any collection of integers $k_1, k_2, \ldots, k_m$, the events $A_1(k_1), \ldots, A_m(k_m)$ are jointly independent;

(ii) $P(N(h) = 1) = \lambda h + o(h)$ as $h \to 0$;

(iii) $P(N(h) \geq 2) = o(h)$ as $h \to 0$.

Our purpose is to derive formulas for the probabilities $P_k(t)$ that there are exactly k calls in the interval of length t.

In addition to the above assumptions (i)–(iii), we postulate that the probability of having more than zero events in a zero-length interval is zero: $P_k(0) = 0,\ k \geq 1$. Therefore, $P_0(0) = 1$.

It follows from Definition 2.1.1 that $P_0(h) = 1 - \lambda h + o(h)$ and thus

$$\begin{aligned} P_0(t+h) &= P(N(t+h) = 0) = P(N(t) = 0, N(t+h) - N(t) = 0) \\ &= P(N(t) = 0)P(N(t+h) - N(t) = 0) \\ &= P_0(t)[1 - \lambda h + o(h)] \ . \end{aligned}$$

Hence,

$$(P_0(t+h) - P_0(t))/h = -\lambda P_0(t) + o(h)/h \ . \tag{2.1.1}$$

Now letting $h \to 0$, we obtain the differential equation

$$P_0'(t) = -\lambda P_0(t), \tag{2.1.2}$$

which has to be solved for the initial condition $P_0(0) = 1$. The solution is

$$P_0(t) = e^{-\lambda t}. \tag{2.1.3}$$

Similarly, for $n > 0$,

$$P_n(t+h) = P_n(t)P_0(h) + P_{n-1}(t)P_1(h) + o(h). \tag{2.1.4}$$

Substitute into $P_0(h) = 1 - \lambda h + o(h)$ in (2.1.4). This leads to the equation

$$P_n'(t) = -\lambda P_n(t) + \lambda P_{n-1}(t) \ . \tag{2.1.5}$$

This must be solved under the condition that $P_n(0) = 0,\ n > 0$. Let us omit the routine solution procedure and provide the answer:

$$P_n(t) = e^{-\lambda t}(\lambda t)^n/n! \ . \tag{2.1.6}$$

(Verify it!) We have therefore proved the following result:

Theorem 2.1.1
In a Poisson process with parameter λ, the number of calls in an interval of length t has a Poisson distribution with parameter λt.

This would be a good time time for a short reminder about the Poisson distribution. We say that the r.v. X has a Poisson distribution with parameter μ (written $X \sim \mathcal{P}(\mu)$) if $P(X = k) = e^{-\mu}\mu^k/k!,\ k = 0, 1, 2, \ldots$. The mean value of X is $E[X] = \mu$ and the variance is $Var[X] = \mu$.

The Poisson distribution is closely related to the *binomial* distribution. Suppose that we have a series of n independent experiments, each of which can be a success (with probability p) or a failure with probability $1-p$. Then the random variable Y which counts the total number of successes in n experiments has the binomial distribution:

$$P(Y = k) = \binom{n}{k} p^k (1-p)^{n-k} . \tag{2.1.7}$$

We will use the notation $Y \sim B(n,p)$.

Now assume that $n \to \infty$ and $p \to 0$, in such a way that $n \cdot p = \mu$. Then it is a matter of simple algebra to show that

$$P(Y = k) = \binom{n}{k} p^k (1-p)^{n-k} \to e^{-\mu}\mu^k/k!, \tag{2.1.8}$$

as $n \to \infty$.

One can imagine the Poisson process with rate λ in the following way. Divide the interval $[0, t]$ into M small intervals, each of length $\Delta = t/M$. Imagine that a "lottery" takes place in each such small interval, the result of which can be either "success" or "failure", independently of the results of all the previous lotteries. Each success is an "event" (a call) in the counting process $N_0(t)$ which is the total number of successes in $[0, t]$. Now let the probability of success be $p_\Delta = \Delta \cdot \lambda$, i.e. proportional to the length of the small interval. Then, as $M \to \infty$,

$$p_\Delta \cdot M = (\lambda \cdot t/M) \cdot M = \lambda t. \tag{2.1.9}$$

The number of successes in $[0, t]$ has, for any finite M, the binomial distribution, which in the limit (as M goes to infinity) approaches the Poisson distribution with parameter λt. The above-described discrete "lottery process" is in fact a discrete version of Definition 2.1.1 (with $\Delta = h$).

Denote by τ_i the interval between the $(i-1)$th and the ith call in the Poisson process. (Assume that the call number zero appears at $t = 0$; see Fig. 2.1.)

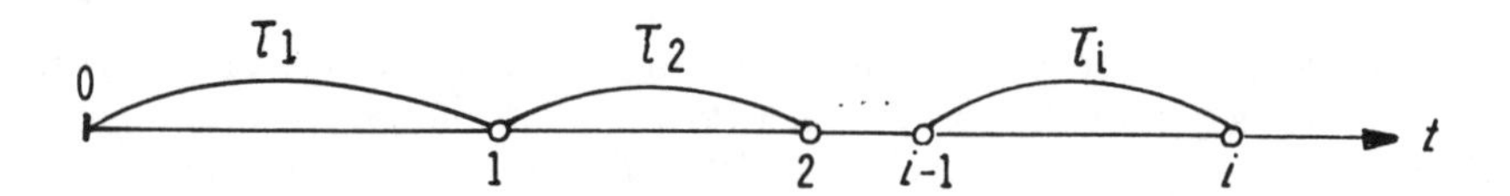

Figure 2.1. Poison process. $\tau_i \sim \text{Exp}(\lambda)$

Theorem 2.1.1 implies the following

Corollary 2.1.1
(i) $\tau_1, \tau_2, \ldots, \tau_n, \ldots$ are i.i.d. random variables,

$$P(\tau_i \le t) = 1 - e^{-\lambda t}. \tag{2.1.10}$$

(ii) Define $T_k = \sum_{i=1}^{k} \tau_i$. Then

$$P(T_k \le t) = 1 - e^{-\lambda t} \sum_{i=1}^{k-1} \frac{(\lambda t)^k}{k!} \,. \tag{2.1.11}$$

The distribution of τ_i is called exponential and will be denoted by $\text{Exp}(\lambda)$. It plays an extremely important role in reliability theory. The distribution of T_k is called the gamma distribution and it will be denoted by $T_k \sim Gamma(k, \lambda)$.

Proof
Part (i) follows from the fact that $P_0(t) = e^{-\lambda t}$. Indeeed, it says that the time to the first call exceeds t with probability $P_0(t)$. Therefore, the time to the first call $\tau_1 \sim \text{Exp}(\lambda)$.

From the definition of the Poisson process it follows that the time τ_2 between the first call and the second call has the same distribution and is *independent* of τ_1, etc.

To prove (ii), note that the events $\{T_k > t\}$ and "there are $(k-1)$ or less events in $[0, T]$" are equivalent. Thus,

$$P(T_k > t) = e^{-\lambda t} \sum_{i=1}^{k-1} \frac{(\lambda t)^k}{k!} \,. \tag{2.1.12}$$

The exponential distribution is a particular case of the gamma distribution: $\text{Exp}(\lambda) = Gamma(1, \lambda)$. The mean and variance of these distributions is given below. For $\tau \sim \text{Exp}(\lambda)$,

$$E[\tau] = 1/\lambda; \;\; Var[\tau] = 1/\lambda^2 \,. \tag{2.1.13}$$

The exponential distribution is the only continuous distribution which possesses the so-called *memoryless property.* Let $\tau \sim \text{Exp}(\lambda)$. Then $P(\tau > t) = \int_t^\infty \lambda e^{-\lambda x} dx = e^{-\lambda t}$. Suppose it is known that $\{\tau > t_1\}$. Let us compute the conditional probability

$$P(\tau > t_1 + t_2 | \tau > t_1) = \frac{P(\tau > t_1 + t_2)}{P(\tau > t_1)} = \frac{e^{-\lambda(t_1+t_2)}}{e^{-\lambda t_1}} = e^{-\lambda t_2}, \qquad (2.1.14)$$

which equals $P(\tau > t_2)$. In physical terms, (2.1.14) says that if the "system" has survived past t_1, the chances of surviving an extra interval of length t_2 are the same as if the system were "brand new".

To get a better understanding of the exponential distribution, let us consider its discrete version. Suppose that the time axis is divided into small intervals of length Δ. A certain event, say a failure, may appear with probability p in each of these intervals, independently of its appearance in any other interval. Let us define the random variable K as the ordinal number of that Δ-interval in which the failure has appeared *for the first time.* Obviously, $P(K = n) = (1-p)^{n-1}p$. It is a matter of routine calculation to show that $P(K > n) = \sum_{i=n+1}^{\infty} P(K = i) = (1-p)^n$. Now let us consider the conditional probability

$$P(K > n+m | K > n) = \frac{P(K > n+m)}{P(K > n)} = (1-p)^m = P(K > m). (2.1.15)$$

This is exactly the discrete analog of the memoryless property of the exponential distribution: if the "system" did not fail during the first n intervals, the chances of not failing at least another m intervals are the same as the chances of not failing during the first m intervals for a "new" system.

A random variable K which is distributed as $P(K = n) = (1-p)^{n-1}p$, $n = 1, 2, 3, \ldots$, is said to have a *geometric* distribution. We will denote this by$K \sim Geom(p)$. For this random variable,

$$E[K] = 1/p; \; Var[K] = (1-p)/p^2 \, . \qquad (2.1.16)$$

The geometric distribution is in fact a discrete version of the exponential distribution. Let us consider a time scale in which the lifetime τ is measured in Δ -units: $\tau = K \cdot \Delta$. Then $P(\tau > t) = P(K\Delta > t) = P(K > [t/\Delta]) = (1-p)^{[t/\Delta]} = (1-p)^{(1/p)\cdot p \cdot [t/\Delta]} \approx e^{-at}$, where $p \approx a \cdot \Delta$ and $p \to 0$. So, if the failure probability is approximately proportional to the interval length and tends to zero, the geometric distribution approaches the exponential distribution.

From the model of the gamma distribution, it follows that

$$E[T_k] = k/\lambda; \; Var[T_k] = k/\lambda^2 \, . \qquad (2.1.17)$$

The densities of the exponential and gamma distributions are, respectively,

$$f_\tau(t) = \lambda e^{-\lambda t}, \; f_{T_k}(t) = \frac{t^k \lambda^{k-1} e^{-\lambda t}}{(k-1)!}, \; t > 0 \, . \qquad (2.1.18)$$

A useful characteristic of a distribution of a positive random variable is the *coefficient of variation* (c.v.) defined as

$$c.v. = \sqrt{Var[\tau]}/E[\tau]\ . \tag{2.1.19}$$

The more peaked is the d.f., the smaller is the c.v. For the gamma family (2.1.18),

$$c.v. = \frac{1}{\sqrt{k}}\ . \tag{2.1.20}$$

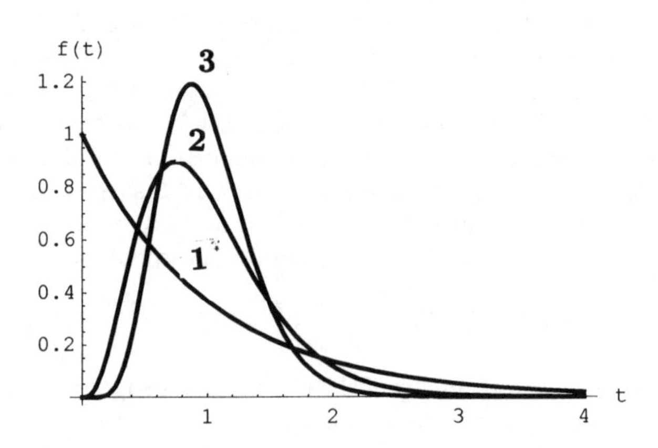

Figure 2.2. The gamma densities. All three densities have the same mean 1. The c.v. is 1, 0.5 and 0.35 for curves 1, 2 and 3, respectively

Figure 2.2 shows the form of the density functions for the gamma family.

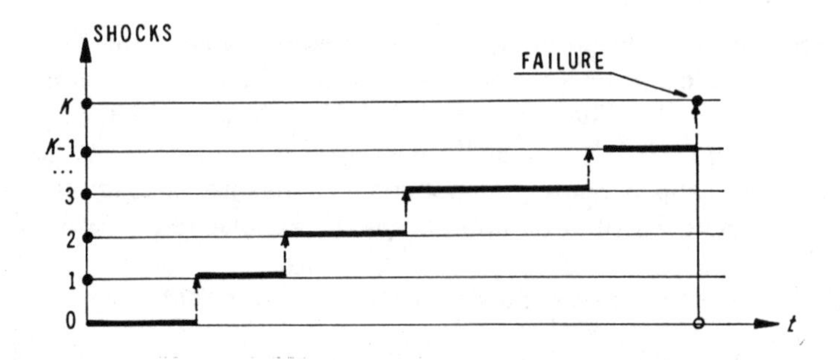

Figure 2.3. The failure appears with the appearance of the kth shock

The gamma distribution is quite useful for reliability modeling because it may describe component (system) lifetime for the so-called shock accumulation scheme. Imagine that external shocks arrive according to a Poisson process with parameter λ and that the component can survive no more than $k-1$ shocks. Then the appearance of the kth shock means failure. Therefore, the component lifetime distribution function will be $Gamma(k, \lambda)$.

For "large" k, a normal approximation can be used and

$$P\Big(\sum_{i=1}^{k} \tau_i \le t\Big) = P\Big(\frac{\sum_{i=1}^{k} \tau_i - k/\lambda}{\sqrt{k}/\lambda} \le \frac{t - k/\lambda}{\sqrt{k}/\lambda}\Big) \approx \Phi\Big(\frac{t-k/\lambda}{\sqrt{k}/\lambda}\Big) \,. \quad (2.1.21)$$

Here $\Phi(\cdot)$ is the standard normal c.d.f. Since the normal random variable formally may have negative values and the lifetime is by definition a positive random variable, the use of normal approximation is justified if its left (negative) tail may be neglected. We recommend using the normal approximation only for $k \ge 9$–12.

A useful modification of (2.1.10) is the exponential distribution with a location parameter. Its density $f(t)$ is 0 for $t < a$, and

$$f(t) = \lambda e^{-\lambda(t-a)}, \quad \text{for } t \ge a. \quad (2.1.22)$$

In applications, a is usually positive, and in reliability theory it is called the *threshold* parameter. The density (2.1.22) might reflect the following single shock failure model. Imagine that that there is some latent deterioration process going on in the system, and during the interval $[0, a-\Delta]$ the deterioration is comparatively small so that the shocks do not cause system failure. During a relatively short time interval $[a-\Delta, a]$, the deterioration progresses rapidly and makes the system susceptible to shocks. Afterwards, a single shock causes failure. The notation $\tau \sim \text{Exp}(a, \lambda)$ is used for the r.v. with d.f. (2.1.22).

Since the exponential distribution plays a prominent role in reliability and preventive maintenance theory, we will present two additional models leading to the exponential distribution. The first is a formalized scheme of a "large" renewable system.

Suppose that the system consists of n independent components, for n "large". The lifetime of component i is τ_i, the mean lifetime of τ_i is $\mu_i = E[\tau_i]$. After the component fails, it is immeadiately replaced by a statistically identical one. Assume that all components are organized in series, and each component failure causes system failure.

Let us consider the ordered sequence of time instants of system failures $t_1 < t_2 < \ldots < t_k < \ldots$. This sequence is obtained by positioning all component failure instants on a common time axis.

The following remarkable fact was first discovered by Khinchine (1956). If the mean component lifetimes are approximately of the same magnitude, then the intervals between system failures $\Delta_m = (t_m - t_{m-1})$ behave, for large m and n, approximately as i.i.d. exponentially distributed r.v.'s. The parameter of these r.v.'s approaches $\Lambda = \sum_{i=1}^{n} \mu_i^{-1}$.

If $\tau_i \sim \text{Exp}(1/\mu_i)$, then the pooled sequence of failures forms a Poisson process with parameter $\Lambda = \sum_{i=1}^{n} \mu_i^{-1}$. This becomes obvious if we recall that in the pooled process, the probability of an event in the interval $[t, t+\delta]$ equals $\delta \sum_{i=1}^{n} \mu_i^{-1} + o(\delta)$. The above described fact about arbitrary sequences of r.v.s τ_i describes a model for creating a random process which approaches, under appropriate conditions, the Poisson process with a constant event rate. For more details see Barlow and Proschan (1975, Sect. 8.4).

Now let us turn to the sum of a *random* number of i.i.d. random variables. Formally, our object will be the random variable

$$S = Y_1 + Y_2 + \ldots + Y_K, \tag{2.1.23}$$

where $K \sim Geom(p)$. It is assumed that K is independent on the i.i.d. r.v.s $Y_1, Y_2, \ldots$. Recall that

$$P(K = r) = (1-p)^{r-1} p, \quad r = 1, 2, 3, \ldots. \tag{2.1.24}$$

The r.v. S is created in the following way. Imagine that we have a coin which shows "heads" (failure) with probability p. We flip the coin until failure appears for the first time. For example, if we obtain the sequence "tails, tails, heads", the failure appears on the third trial. The r.v. K counts the number of trials needed to obtain heads for the first time. After K becomes known and $K = k$ is observed, we put $S = Y_1 + \ldots + Y_k$.

Let us give a practical interpretation of (2.1.23). Suppose that the system lifetime is described by an r.v. Y_i. At the instant of failure, an emergency unit replaces the failed system. It is capable of working only a short period of time, say 1 hour. During this time the failed system must be diagnosed and repaired. Let $1 - p$ be the probability that this mission could be carried out with success. If the repair succeeds, the emergency unit returns to standby, and the system continues to work. Otherwise, with probability p, the system fails. One hour is considered a negligible period of time.

An interesting fact is the following:

Theorem 2.1.2
If $Y_i \sim \text{Exp}(1/\mu)$, then

$$S \sim \text{Exp}(p/\mu), \tag{2.1.25}$$

i.e. S has an *exact* exponential distribution with parameter p/μ.
We leave the proof as an exercise.

What happens if Y_i is not an exponential r.v.? It turns out that if p is "small" then S is "close" to the exponential distribution with parameter $p/E[Y_i]$. For more details we refer to Gertsbakh (1989, Chap. 3).

2.2 Aging and Failure Rate

Let us consider in this section a system (component) which is not renewed at its failure. The event "failure" might be defined in many ways, depending on the appropriate circumstances. It might be a mechanical breakdown, deterioration beyond the permissible value, appearance of certain defect in the system performance (overheating, noise), or a decrease in system performance index below a certain previously established critical level.

For our formal treatment, the particular definition of the failure is not important, but what is important is that we fix a certain time scale, define the event "failure" and measure the time τ elapsed from the system's "birth" (the beginning of its operation) to its "death" or failure. We assume further that the ***lifetime***, i.e. the time from "birth" to "death", is a continuous positive random variable with density function $f(t)$ and c.d.f. $F(t) = P(\tau \le t)$.

In real life, on failure, the system is usually repaired or restored, partially or completely, and starts a "new life", fails again, etc. In other words, most systems are subject to renewal in the course of their operation. Our analysis in this section deals with nonrenewable systems: we study the properties of the time interval until the appearance of the first failure.

Of considerable importance and use in reliability analysis is the conditional probability of system failure in a small interval $[t, t + \Delta]$ ***given*** that at time t the system is "up" ("alive"). Formally, we are interested in the conditional probability

$$P(t \le \tau \le t + \Delta | \tau > t) = \frac{P(\text{failure appears in } (t, t + \Delta))}{P(\tau > t)} . \tag{2.2.1}$$

Since the numerator in (2.2.1) equals approximately $f(t) \cdot \Delta$, we arrive at the following formula:

$$P(t \le \tau \le t + \Delta | \tau > t) \approx \frac{f(t) \cdot \Delta}{1 - F(t)} . \tag{2.2.2}$$

In other words, the conditional failure probability in $[t, t + \Delta]$ given the system is up at t is, for small Δ, approximately ***proportional*** to $f(t)/(1 - F(t))$.

Definition 2.2.1

Suppose that the lifetime is a positive random variable with d.f. $f(t)$ and c.d.f. $F(t)$. Then the function

$$h(t) = \frac{f(t)}{1 - F(t)} \tag{2.2.3}$$

is called the ***failure or hazard rate***. Note that $h(t)$ is defined only for t such that $F(t) < 1$.

In the branch of survival analysis which deals with the probabilistic analysis of lifetime for biological objects, the notion of ***rate of mortality*** is used instead of the "failure rate."

Recalling that the exponential distribution arises as the time to the first event in the Poisson process, and that the probability of the appearance of an event in $[t, t+\Delta]$ is proportional to Δ, it should be expected that the failure rate for the exponentially distributed lifetime is constant. Indeed, for $f(t) = \lambda e^{-\lambda t}$ and $F(t) = 1 - e^{-\lambda t}$,

$$h(t) = \frac{f(t)}{1 - F(t)} = \lambda . \tag{2.2.4}$$

Suppose that we know the system failure rate $h(t)$. Could we reconstruct in a unique way its lifetime density function and/or the lifetime c.d.f.? The answer is "yes" since there is a simple and useful relationship between the survival probability $R(t) = 1 - F(t) = P(\tau > t)$ and the failure rate $h(t)$:

$$P(\tau > t) = R(t) = \exp\Big(- \int_0^t h(u)du \Big) . \tag{2.2.5}$$

To prove it represent the integral in (2.2.5) as $- \int_0^t d[\log(1 - F(t)]$.

Definition 2.2.2
A random variable τ is of increasing (decreasing) failure rate type if the corresponding failure rate $h(t)$ is an increasing (decreasing) function of t.

We write $\tau \in$ IFR (DFR) or $F \in$ IFR (DFR) if $h(t)$ is increasing (decreasing).

The exponential random variable is, by definition, both of IFR and DFR type.

A phenomenon which is very important, for reliability theory in general and for preventive maintenance in particular, is *aging*. On an intuitive level, aging means an increase of failure risk as a function of time in use. In that case the failure rate is an appropriate measure of aging. The memoryless exponential distribution reflects the "no aging" situation. A lifetime $\tau \in$ IFR displays an increasing failure risk and reflects the situation when the object is subject to aging. If $\tau \in$ DFR, the failure risk decreases in time, and the corresponding system, contrary to aging, becomes "younger." We will show later that mixtures of exponential distributions have this somewhat surprising property.

Another approach to the formalization of aging is considering the conditional survival probability $R(x|t)$, as a function of t, for an interval of fixed length x, $[t, t + x]$, *given* that the system is alive at t. Obviously,

$$R(x|t) = \frac{P(\tau > t + x)}{P(\tau > t)} = \frac{R(t + x)}{R(t)} . \tag{2.2.6}$$

If this conditional probability decreases as a function of t, it would be natural to say that the system is aging. It turns out that this type of aging is completely identical to the IFR property. Indeed, using (2.2.5), it is easy to establish that $R(x|t) = \exp(- \int_t^{t+x} h(u)du)$. Taking the derivative of $R(x|t)$ with respect to t, we arrive at the following theorem:

Theorem 2.2.2
$R(x|t)$ is decreasing (increasing) in t if $h(t)$ is increasing (decreasing).

In the case of discrete random variables, the definition of IFR and DFR should be given in the terms of the survival probability $R(x|t)$.

Definition 2.2.3
Let X_i, $i = 1, \ldots, n$, be independent random variables with d.f. $f_i(t)$. We say that Y is a *mixture* of $X_1, \ldots, X_n$ if its d.f. is

$$f_Y(t) = \sum_{i=1}^n \alpha_i f_i(t), \text{ where } 0 \le \alpha_i \le 1, \ \sum_{i=1}^n \alpha_i = 1.$$

Mixtures arise, for example, if the daily production of several similar machines is mixed together. α_i is then the relative portion of the ith machine in the daily production; X_i is the random characteristic of the items produced on the ith machine, say the random lifetime of a resistor which came out of the ith machine. An interesting fact which also has important practical implications is stated in the following

Theorem 2.2.3.
A mixture of exponential distributions has a DFR distribution.

Proof
Let $X_i \sim \text{Exp}(\lambda_i)$. Obviously, the probability $P(Y > t)$ can be represented in the form $P(Y > t) = E[e^{-\Lambda t}]$, where Λ is a random variable such that $P(\Lambda = \lambda_i) = \alpha_i$. Then the failure rate of Y is

$$h_Y(t) = \frac{-\partial E[e^{-\Lambda t}]}{\partial t} / E[e^{-\Lambda t}] \, .$$

Now consider the derivative of $h_Y(t)$ with respect to t. The differentiation can be carried out inside the $E[\cdot]$ operator:

$$\frac{dh_Y(t)}{dt} = \frac{-E[e^{\Lambda t}]E[\Lambda^2 e^{\Lambda t}] + (E[\Lambda e^{\Lambda t}])^2}{(E[e^{\Lambda t}])^2} \, .$$

Now define $X = e^{\Lambda t/2}$ and $Z = \Lambda e^{-\Lambda t/2}$. Then the numerator in the last expression is $E^2[XZ] - E[X^2]E[Z^2]$. It is nonpositive by the Cauchy–Schwarz inequality, which proves the theorem.

Let us now investigate the relationship between the failure rate of components and the failure rate of the system. It is assumed that the system consists of independent components. The situation is very simple for a series system, as the following theorem states:

Theorem 2.2.4
The failure rate of a series system is the sum of its component failure rates.

Proof
The reliability of a series system is $R(t) = \prod_{i=1}^{n} R_i(t)$. Take the logarithm of both sides. Differentiate with respect to t. Then we obtain that

$$h(t) = \sum_{i=1}^{n} h_i(t), \tag{2.2.7}$$

where $h_i(t) = f_i(t)/R_i(t)$.

It is clear from (2.2.7) that if all components are of IFR (DFR) type, then the series system is of the IFR (DFR) type.

Historically, the fact established in Theorem 2.2.4 became an important impulse for the design of reliable systems and for the development of reliability theory. During the era of first-generation computers, designers and users suddenly became aware that their computers fail extremely often, say once per hour. They realized that it was because the computer consisted of tens of thousands of parts (tubes, relays, etc.) which had failure rates of the order of 10^{-5} hr^{-1}. As a system, the computer is basically a series system, i.e. the failure of each part means the failure of the system. Thus the system failure rate for the first-generation computers was of the order of $1 - 0.5$ hr^{-1}. Thus the average interval between failures was 1–2 hours.

There is an interesting and useful connection between the mean lifetimes of the components of a series system and the mean system lifetime. We state it without proof.

Theorem 2.2.5
Let μ_i be the mean lifetime of the ith component. If all components are of IFR type, then the system mean lifetime μ satisfies the following inequality:

$$\mu > \Big(\sum_{i=1}^{n} \mu_i^{-1}\Big)^{-1}. \tag{2.2.8}$$

We have seen that the IFR property is inherited by a series system from its components. The situation differs for a *parallel* system. It may happen that the IFR property of the components in *not* inherited by the whole system. Exercise

3 demonstrates that a system of two parallel elements with exponential lifetimes may have a *nonmonotone* failure rate.

The following theorem shows that the exponential distribution represents the "worst case" from the reliability point of view. Denote by τ the system lifetime.

Theorem 2.2.6
If $F \in$ IFR and $E[\tau] = \mu$, then, for $0 \le t \le \mu$,

$$R(t) \ge e^{-t/\mu} . \tag{2.2.9}$$

We omit the proof, which can be found in Gertsbakh (1989, p. 63). This theorem has an immediate and very useful implication.

Corollary 2.2.6
Let $r_0 = \psi(p_1, p_2, \ldots, p_n)$ be the system reliability function. Then, for $0 \le t \le \min\{\mu_1, \mu_2, \ldots, \mu_n\}$,

$$R(t) \ge \psi(e^{-t/\mu_1}, e^{-t/\mu_2}, \ldots, e^{-t/\mu_n}) . \tag{2.2.10}$$

Proof
This follows immediately from Theorem 2.2.6 and the monotone dependence of $\psi(p_1, \ldots, p_n)$ on each p_i.

This statement says that on the time interval $[0, \min\{\mu_1, \mu_2, \ldots, \mu_n\}]$, the worst-case situation would take place if all system components (being organized in a system with the same structure function) were replaced by exponential ones, with the same mean lifetimes.

We can view the results of the last two theorems as an attempt to extract information about the system lifetime when the information available is the mean lifetime and the behavior of the failure rate. The more we know about the statistical properties of the system, the more accurately we can predict system lifetime. An elegant mathematical theory was developed by Barlow and Marshall (1964) for IFR distributions with known ***first*** and ***second*** moment. This theory allows us to obtain a quite accurate ***bounds*** on system reliability. This turns out to be very useful for various applications, including preventive maintenance.

2.3 Normal, Lognormal and Weibull Families

2.3.1 Normal and Lognormal Distributions

We have already introduced the normal distribution in the previous section in the context of a damage accumulation process described by the model $\tau =$

$\tau_1 + \tau_2 + \ldots + \tau_k$, where τ_i are the intervals between the successive events in Poisson process. We say that $\tau \sim N(\mu, \sigma)$ if

$$P(\tau \leq t) = \Phi\Big(\frac{t-\mu}{\sigma}\Big) = \frac{1}{\sqrt{2\pi}} \int_{-\infty}^{(t-\mu)/\sigma} \exp[-v^2/2]dv \ . \qquad (2.3.1)$$

μ and σ are the mean value and the standard deviation of the r.v. τ, respectively. The support of the normal distribution is the whole axis $(-\infty, \infty)$ and therefore formally this distribution cannot represent a positive random variable. However, if $\sigma/\mu < 1/3$, the negative tail of the normal distribution is negligible, and it may serve as a c.d.f. for a positive random variable. Let us compare, for example, the normal and gamma distributions for $k = 10$. By (2.1.20), the c.v. of the gamma distribution is $1/\sqrt{10}$. Thus we might expect the survival probabilities
$R_1(t) = e^{-\lambda t} \sum_{i=0}^{9} \lambda^i/i!$ and $R_2 = 1 - \Phi\Big((t - 10/\lambda)/\sqrt{10}\lambda^{-1}\Big)$
to be close to each other (both distributions have the same mean and variance). This is in fact true, as Fig. 2.4 shows.

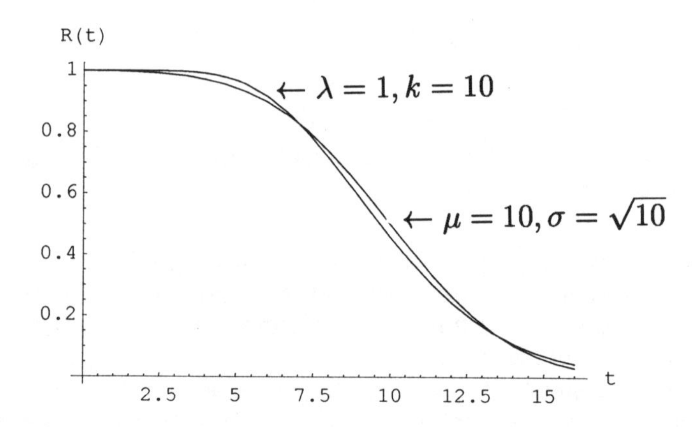

Figure 2.4. Comparison of gamma and normal survival probabilities

The normal random variable is of the IFR type. Figure 2.5 shows typical curves of the failure rate.

The *lognormal* distribution is defined in the following way. We say that the r.v. Y has a lognormal distribution with parameters μ and σ (which will be denoted as $Y \sim \log N(\mu, \sigma)$), if

$$P(Y \leq t) = \Phi\Big(\frac{\log t - \mu}{\sigma}\Big). \qquad (2.3.2)$$

The corresponding density function is

$$f_Y(t; \mu, \sigma) = \frac{1}{\sqrt{2\pi}\sigma t} \exp\Big(-\frac{(\log t - \mu)^2}{2\sigma^2}\Big), \text{ for } t > 0 \ . \qquad (2.3.3)$$

Let $\tau = \log Y$. How is τ distributed?

$$P(\tau \leq t) = P(\log Y \leq t) = P(Y \leq e^t) = \Phi\Big(\frac{t-\mu}{\sigma}\Big). \tag{2.3.4}$$

We see, therefore, that $\tau = \log Y \sim N(\mu, \sigma)$. The logarithm of a lognormal r.v. is a normal random variable.

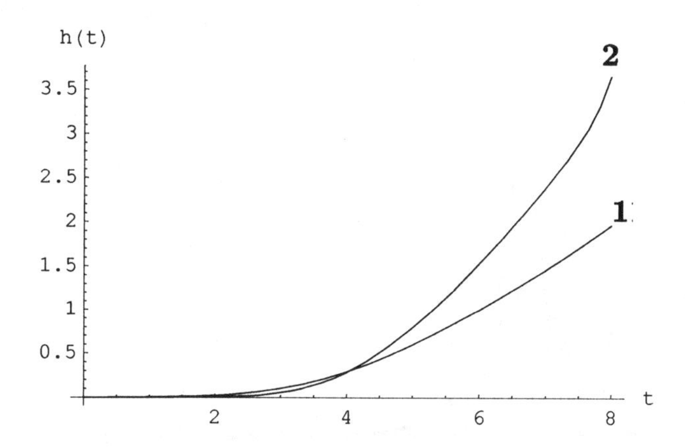

Figure 2.5. $h(t)$ for the normal distribution. 1 – for $N(\mu = 5, \sigma = 1)$, 2 – for $N(\mu = 5, \sigma = 1.33)$

The first two central moments of the lognormal distribution are as follows:

$$E[Y] = \exp(\mu + \sigma^2/2), \tag{2.3.5}$$

$$Var[Y] = (e^{\sigma^2} - 1) \cdot \exp(2\mu + \sigma^2). \tag{2.3.6}$$

For the lognormal distribution, the coefficient of variation is a function of σ only:

$$c.v. = \sqrt{e^{\sigma^2} - 1} \approx \sigma \ \text{ for } \sigma < 0.5\,. \tag{2.3.7}$$

Table 2.1 shows how the c.v. depends on σ.

The lognormal density is left-skewed and has for large values of c.v. a heavy right tail. The density becomes more symmetrical for small c.v. values; see Fig. 2.6.

The *three-parameter* lognormal distribution has density

$$f_Y(t; \mu, \sigma) = \frac{1}{\sqrt{2\pi}\sigma(t - t_0)} \exp\Big(-\frac{(\log(t - t_0) - \mu)^2}{2\sigma^2}\Big), \text{ for } t > t_0, \tag{2.3.8}$$

and t_0 is called the threshold of sensitivity. The density (2.3.8) is used when the lifetime data follow the lognormal density for $t > t_0$, and failures were not observed on $[0, t_0]$.

Table 2.1: Coefficient of variation for the lognormal distribution

σ	0.1	0.15	0.2	0.25	0.3	0.4	0.5	0.6	0.7
$c.v.$	0.10	0.151	0.202	0.254	0.307	0.416	0.533	0.658	0.795

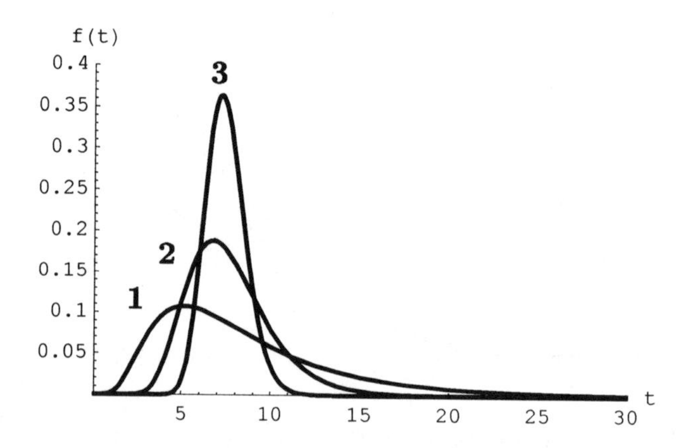

Figure 2.6. The lognormal density for $\mu = 2, \sigma = 0.6$ (curve 1), $\mu = 2, \sigma = 0.3$ (curve 2) and $\mu = 2, \sigma = 0.15$ (curve 3)

Investigation of the failure rate $h(t) = f_Y(t;\mu,\sigma)/(1 - \Phi((\log t - \mu)/\sigma))$ reveals that $h(t) \to 0$ as $t \to 0$ and as $t \to \infty$. Thus, the lognormal random variable is neither of IFR nor of DFR type. This distribution may be suitable for describing the lifetime of objects which display so-called "training effects;" see Gertsbakh and Kordonsky (1969, p. 76) for more details.

There are several probabilistic models in the reliability literature describing the appearance of the lognormal distribution. An interesting source is the book of Aitchison and Brown (1957). Gertsbakh and Kordonsky (1969) show that the lognormal distibution arises in a model of damage accumulation in which the probability of a single damage occurrence in $[t, t+\Delta]$ is $\Delta \cdot \lambda/(1+t)$. In fact, the lifetime defined as the time to the appearance of the kth damage event is the time to the kth event in a nonhomogeneous Poisson process with intensity function $\lambda(t) = \lambda/(1+t)$; see Appendix A.

2.3.2 The Weibull distribution

Lawless (1983) writes that about half of all papers on statistical methods in reliability in the period 1967–1980 were devoted to the Weibull distribution. The popularity of this distribution lies in the fact that, depending on the parameters, it may describe both IFR and DFR random variables. Besides, a logarithmic transformation of the Weibull random variable produces an r.v. which belongs to the so-called location–scale family which possesses several very good features for statistical analysis.

We say that the r.v. τ has a Weibull distribution with parameters λ and β if the density function of τ equals, for $t > 0$,

$$f(t; \lambda, \beta) = \lambda^\beta \beta t^{\beta-1} \exp(-(\lambda t)^\beta), \tag{2.3.9}$$

and zero otherwise. $\lambda > 0$ is called the scale parameter, and $\beta > 0$ the shape parameter.

The Weibull c.d.f. is

$$F(t; \lambda, \beta) = 1 - \exp(-(\lambda t)^\beta) \text{ for } t > 0, \tag{2.3.10}$$

and $F(t; \lambda, \beta) = 0$ for $t \leq 0$. The notation $\tau \sim W(\lambda, \beta)$ will be used for the Weibull distribution.

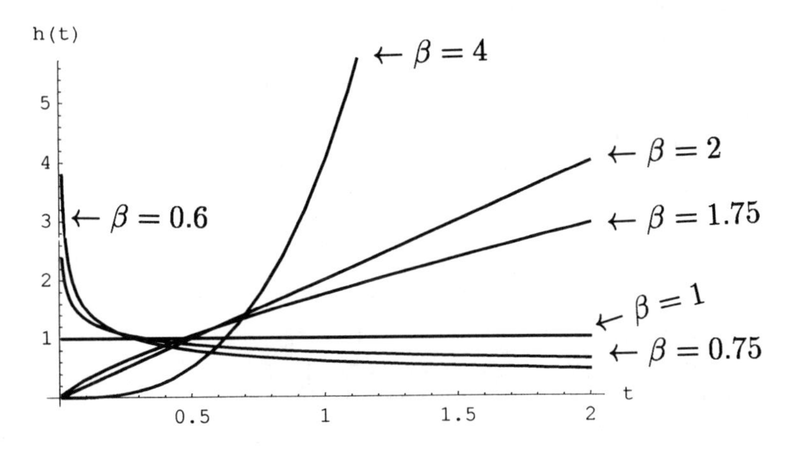

Figure 2.7. $h(t)$ curves for the Weibull distribution ($\lambda = 1$)

Obviously, for $\beta = 1$, the Weibull distribution equates to the exponential distribution. The failure rate for the Weibull family is

$$h(t) = f(t; \lambda, \beta)/(1 - F(t; \lambda, \beta)) = \lambda^\beta \beta t^{\beta-1}. \tag{2.3.11}$$

$h(t)$ is increasing in t for $\beta > 1$, decreasing in t for $\beta < 1$ and remains constant for $\beta = 1$; see Fig. 2.7.

The mean value of the Weibull random variable is

$$E[\tau] = \lambda^{-1}\Gamma(1 + \beta^{-1}), \tag{2.3.12}$$

where $\Gamma(\cdot)$ is the gamma function: $\Gamma(x) = \int_0^\infty v^{x-1}e^{-v}dv$. The variance is also expressed via the gamma function:

$$Var[\tau] = \lambda^{-2}[\Gamma(1 + 2\beta^{-1}) - (\Gamma(1 + \beta^{-1}))^2]\,. \tag{2.3.13}$$

Table 2.2: The values of $\Gamma[1+1/\beta]$, $\Gamma[1+2/\beta]$ and $\Gamma[1+2/\beta]-(\Gamma[1+1/\beta])^2$

β	$\Gamma[1+1/\beta]$	$\Gamma[1+2/\beta]$	$\Gamma[1+2/\beta]-(\Gamma[1+1/\beta])^2$
0.75	1.19064	4.01220	2.59458
1.00	1.00000	2.00000	1.00000
1.25	0.93138	1.42962	0.56215
1.50	0.90274	1.19064	0.37569
1.75	0.89062	1.06907	0.27587
2.00	0.88623	1.00000	0.21460
2.25	0.88573	0.95801	0.17349
2.50	0.88726	0.93138	0.14415
2.75	0.88986	0.91409	0.12224
3.00	0.89298	0.90275	0.10533
3.25	0.89633	0.89534	0.09193
3.50	0.89975	0.89062	0.08107
3.75	0.90312	0.88776	0.07213
4.00	0.90640	0.88623	0.06466
4.25	0.90956	0.88564	0.05834
4.50	0.91257	0.88573	0.05294
4.75	0.91544	0.88632	0.04828
5.00	0.91817	0.88726	0.04423

Table 2.3: The c.v. for the Weibull family

β	0.5	1	1.5	2	3	4	5
$c.v.$	2.24	1	0.68	0.52	0.36	0.28	0.23

To make it easier to calculate the mean value and the variance, Table 2.2 gives the values of $\Gamma[1+1/\beta]$, $\Gamma[1+2/\beta]$ and $\Gamma[1+2/\beta]-(\Gamma[1+1/\beta])^2$ for various β values.

For the Weibull distribution the c.v. depends only on β:

$$c.v. = \Big(\Gamma(1+2/\beta)/\Gamma^2(1+1/\beta) - 1\Big)^{0.5}. \tag{2.3.14}$$

Table 2.3 shows how this dependence for a few β values.

The Weibull densities for various c.v. values are shown in Fig. 2.8.

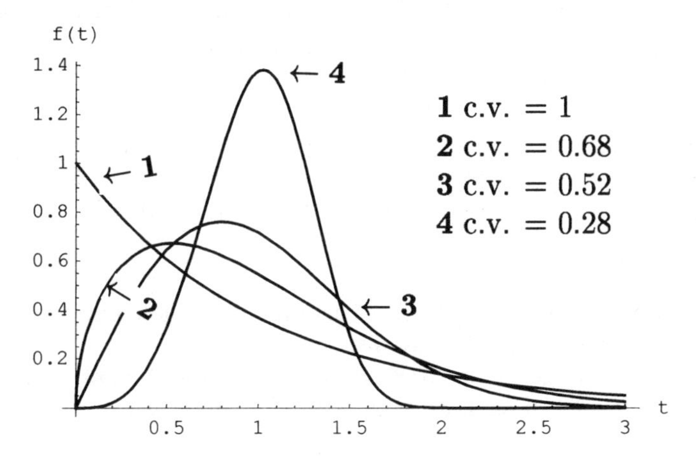

Figure 2.8. The Weibull densities (mean = 1)

Let $\tau \sim W(\lambda, \beta)$. Consider the random variable $X = \log \tau$. It has the following c.d.f.:

$$P(X \le t) = F(t) = 1 - \exp^{-\exp((t-a)/b)}, \quad -\infty < t < \infty, \tag{2.3.15}$$

where $a = -\log \lambda$ and $b = 1/\beta$. The derivation of (2.3.15) is left as an exercise.

The c.d.f. (2.3.15) is of location–scale type. This fact is very important for statistical inference about the location and scale parameters.

Definition 2.3.1

Suppose that the r.v. X has a c.d.f. $F(t; \mu, \sigma)$, where $\sigma > 0$, and that the c.d.f. $F_0(t)$ is parameter-free. We say that $F(t; \mu, \sigma)$ is of location–scale type if $F(t; \mu, \sigma) = F_0((t - \mu)/\sigma)$.

The corresponding density function is $f_0((t-\mu)/\sigma) \cdot \sigma^{-1}$; μ and σ are the location and scale parameters, respectively.

The two-parameter exponential distribution (2.1.22), the normal and the extreme-value distribution (2.3.15) belong to location–scale families.

The notation $X \sim \text{Extr}(a, b)$ will be used to denote an r.v. with c.d.f. (2.3.15). This distribution is known in probability theory as the extreme-value distribution of the third type. For more details see Barlow and Proschan (1975, Chap. 8).

Suppose that we have a collection of independent r.v.'s $\tau_1, \ldots, \tau_n$, such that the corresponding c.d.f.'s behave near zero as $P(\tau_k \le t) = ct^d(1 + o(1))$, $c > 0$, $d > 0$; $t > 0$. Then consider the random variable $Z_n = a_n \min(\tau_1, \ldots, \tau_n)$, where a_n is a normalizing constant, $a_n = n^{1/d}$. Then it can be proved that Z_n converges in distribution to the Weibull random variable, with parameters $c^{1/d}$ and d, as n tends to infinity.

The Weibull family is closed with respect to the minimum-type operation, in the following sense. If $\tau_i \sim W(\lambda_i, \beta)$, τ_i are independent r.v.'s and $\tau_{(1)} = \min_{1 \le i \le n}\{\tau_i\}$, then $\tau_{(1)} \sim W(\lambda_0, \beta)$, where $\lambda_0 = (\sum_{i=1}^n \lambda_i^\beta)^{1/\beta}$. We leave the reader to prove this property as an exercise.

Let us consider an arbitrary monotone system with independent and nonrenewable components. The component i has lifetime $\tau_i \sim \text{Exp}(\lambda_i)$. Let us assume that the components are "highly reliable", i.e. formally

$$\lambda_i = \alpha \cdot \theta_i, \quad \alpha \to 0. \tag{2.3.16}$$

Following Burtin and Pittel (1972), let us show that the lifetime of this system can be approximated by a Weibull distribution.

For any state vector $\mathbf{x}$, let $G(\mathbf{x}) = \{i : x_i = 1\}$ and $B(\mathbf{x}) = \{i : x_i = 0\}$. Denote by D the set of all state vectors which correspond to the system down state, i.e.

$$D = \{\mathbf{x} : \phi(\mathbf{x}) = \mathbf{0}\}. \tag{2.3.17}$$

The set D can be partitioned into subsets D_r, $D_{r+1}, \ldots$, according to the number of failed elements in the state vectors: $D = \bigcup_{k=r}^n D_k$. So, D_r is a collection of all state vectors which have the ***smallest*** number r of failed elements. (In other words, r is the size of the smallest cut set in the system.)

Let τ be the system lifetime. Let us partition the event "the system is down" into the events "the system is in D_k," $k \ge r$. Then

$$R(t) = 1 - \sum_{k=r}^{n} \Big(\sum_{\mathbf{x} \in D_k} \prod_{\{i \in G(\mathbf{x}),\, \mathbf{x} \in D_k\}} e^{-\lambda_i t} \prod_{\{i \in B(\mathbf{x}),\, \mathbf{x} \in D_k\}} (1 - e^{-\lambda_i t}) \Big). \tag{2.3.18}$$

Indeed, the sum in (2.3.18) is the probability that the system is in one of its down states. For each such state $\mathbf{x}$, the elements of $G(\mathbf{x})$ are up and the elements of $B(\mathbf{x})$ are down. Now expand $e^{-\lambda_i t} = 1 - \lambda_i t + O(\lambda_i^2 t^2) = 1 - \alpha\theta_i t + O(\alpha^2)$ and substitute it into (2.3.18).

The main term in (2.3.18) will be determined by the first summand with the smallest $k = r$. After some algebra it follows that

$$R(t) = P(\tau > t) = 1 - \alpha^r t^r g(\theta) + O(\alpha^{r+1}) \approx \exp(-\alpha^r t^r g(\theta)), \tag{2.3.19}$$

where $g(\cdot)$ is the sum of the products of θ_i over all cut sets of minimal size r:

$$g(\theta) = \sum_{\mathbf{x} \in D_r} \prod_{i \in B(\mathbf{x})} \theta_i. \tag{2.3.20}$$

Equation (2.3.19) suggests that the lifetime of a system with highly reliable exponential component lives approximately follows the Weibull distribution:

$$P(\tau \le t) \approx 1 - \exp(-\alpha^r t^r g(\theta)). \tag{2.3.21}$$

Remark
More formally, (2.3.21) states that for any t in a *fixed* interval $[0, t_0]$, $R(t)$ approaches $\exp(-\alpha^r t^r g(\theta))$ as $\alpha \to 0$.

If there are reasons to believe that the lifetime follows the Weibull pattern only for $t \geq t_0 > 0$ and that failure can never occur before t_0, the natural extension of (2.3.10) is the three-parameter Weibull distribution:

$$P(\tau \leq t) = 1 - \exp(-(\lambda(t - t_0))^\beta),\ t \geq t_0. \tag{2.3.22}$$

As in the lognormal case, t_0 is called a threshold or a guaranteed time parameter.

Let us consider an example illustrating the quality of the approximation (2.3.21).

Example 2.3.1: s–t connectivity of a dodecahedron network
Figure 2.9 shows a network with 20 nodes and 30 edges called dodecahedron. The nodes are absolutely reliable. The edges fail independently and have lifetimes $\tau \sim \text{Exp}(\lambda)$. The network fails if there is no path leading from node 1 ("source") to node 2 ("terminal"). The reliability of such a network is termed the s–t connectivity.

We assume that all $\theta_i = 1$ and α is "small." The dodecahedron has two minimal-size minimal cut sets with $r = 3$ edges. Indeed, node 1 is disconnected from node 2 if the edges $\{17, 18, 19\}$ or the edges $\{6, 19, 30\}$ fail. All other cut sets separating the source from the terminal have size greater than $r = 3$

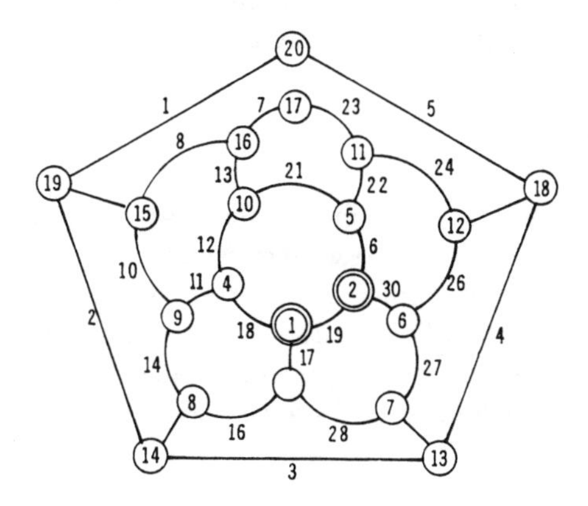

Figure 2.9. The dodecahedron network

Let us take $t = 1$ and consider how good is the approximation to network failure probability provided by the expression $F_{approx}(1) = 1 - \exp[\alpha^3 g(\theta)]$, for

Table 2.4: Comparison of exact and approximate reliability

$e^{-\alpha}$	α	$F_{approx}(1)$	$F_{exact}(1)$	rel. error (%)
0.8	0.223 14	0.021 9	0.036 2	39
0.85	0.162 52	0.085 4	0.012 2	30
0.90	0.105 36	0.002 34	0.002 88	18
0.92	0.083 38	0.001 16	0.001 36	15
0.94	0.061 88	0.000 474	0.000 528	10
0.96	0.040 82	0.000 136	0.000 146	7
0.98	0.020 20	1.65×10^{-5}	1.7×10^{-5}	3
0.99	0.010 05	2.0×10^{-6}	2.03×10^{-6}	1.5

various values of α approaching zero. Since there are two minimal cuts of size 3, $g(\theta) = 2\theta^3 = 2$ by (2.3.20). Thus our approximation is $F_{approx}(1) = 1 - \exp[2\alpha^3]$.

Table 2.4 shows the values of network reliability using the Burtin–Pittel approximation versus the exact values of network failure probabilities $F_{exact}(1)$, for α ranging from $0.22314 = -\ln 0.8$ to $0.01005 = -\ln 0.99$. (Values of $F_{exact}(1)$ were computed by J.S. Provan using an algorithm based on cutset enumeration; see Fishman 1996, p. 62).

It is seen from Table 2.4 that for α below 0.05 the Burtin–Pittel approximation is quite satisfactory.

Estimation of network reliability may be a quite difficult task, especially for large and highly reliable networks. The exact calculation needs special algorithms and software. Monte Carlo simulation is a good alternative, see e.g. Elperin et al (1991). Example 1.3.1 shows that for very reliable networks with failure probability less than 10^{-4}, the Burtin–Pittel approximation provides a reasonably accurate solution with minimal efforts.

2.4 Exercises

1. A system of five identical components operates in the following way. One component is operating while the rest four remain idle. When the operating component fails, its place is immediately taken by an idle unit. The failed component is not renewed. (This situation is often termed "cold standby.") System failure occurs when all components have failed. Assume that component life is exponential with parameter λ. Deduce that system lifetime is Gamma(5, λ).

2 Let $\tau \sim U(0,1)$. Find the failure rate $h(t)$.

3. Two exponential components with failure rates $\lambda_1 = 1$ and $\lambda_2 = 5$ are connected in parallel. Find the expression for system failure rate. Investigate theoretically and/or numerically the behavior of the system failure rate. Is it monotone?

4. Assume that at time $t = 0$ we start testing n identical and independent devices, each having an exponential lifetime with parameter λ. Each failed device is not replaced, and we stop after observing r failures, $r \le n$. Denote by TTT the *total time on test*. Prove that TTT $\sim$ *Gamma*(r, λ).
Hint. The time to the first failure is $\sim$ Exp$(n\lambda)$. The time between the first and the second failure is $\sim$ Exp$((n-1)\lambda)$. Then note that TTT $= \sum_{i=1}^{r}(n - i + 1)(\tau_{(i)} - \tau_{(i-1)})$.

5. Suppose that a population with d.f. Exp$(\lambda_1 = 1)$ is contaminated by adding some amount of units with lifetime distributed according to Exp$(\lambda_2) = 5$. The lifetime of a randomly chosen unit is distributed according to $F(t) = 0.2[1 - e^{-5t}] + 0.8[1 - e^{-t}]$.

Investigate the failure rate theoretically or numerically and show that it is of the DFR type.

6. Suppose that $\tau \sim$ W(λ, β), $E[\tau_1] = 1, Var[\tau_2] = 0.16$. Find λ and β.

Repeat the same calculations by assuming that τ has a lognormal distribution.

7. Element a is connected in series to a parallel group of two identical elements b and c. $\tau_a \sim W(\lambda = 1, \beta = 2)$. Elements b and c are exponential with mean lifetime 0.5. Find the analytic form of the system failure rate $h(t)$.

8. For the system in Exercise 7, use Corollary 2.2.6 to find the lower bound on the system reliability. What is the interval for which this lower bound is valid?

9. Prove formula (2.2.5).

10. Prove Theorem 2.1.2.
Hint: Use the Laplace transform. The Laplace transform of the exponential density is

$L(s) = E[e^{-st}] = \int_0^\infty e^{-st}\lambda e^{-\lambda t}dt = \lambda/(\lambda + s)$.

11. Let $\tau \sim W(\lambda, \beta)$. Find the c.d.f. of $X = \log \tau$. Introduce new parameters $b = \beta^{-1}$ and $a = -\log \lambda$.

12. Let $\tau_i \sim W(\lambda_i, \beta)$. Find the c.d.f. of $\tau = \min(\tau_1, \ldots, \tau_n)$.

13. Show that the Burtin–Pittel approximation for the system lifetime of the bridge system shown on Fig. 1.5 is given by $g(\theta) = \theta_1\theta_2 + \theta_4\theta_5$.

14. Consider a series system of independent components with failure rates $h_1(t), \ldots, h_n(t)$. Prove that the mean lifetime for this system is given by $E[\tau] = \int_0^\infty \exp[-\int_0^t \sum_{i=1}^n h_i(x)dx]dt$.
Hint. Use (1.3.12), (2.2.5) and Theorem 2.2.4.

15a. Find the c.d.f. of a series system consisting of two independent components. The lifetime of the first one is $\tau_1 \sim \text{Exp}(\lambda_1)$ and the lifetime of the second one is $\tau_2 \sim W(\lambda_2 = 0.906, \beta = 4)$.

b. Find the density function of the system lifetime and investigate it graphically for λ_1 values $0.5, 1, 2$ and 4.

Chapter 3

Statistical Inference from Incomplete Data

Statistics is the science of producing unreliable facts from reliable figures.
Quips and Quotes, p. 765.

He uses statistics as a drunkard uses a lamppost, for support, not for illumination.
Chesterton

3.1 Kaplan–Meier Estimator of the Reliability Function

Central to the statistical inference in reliability theory is the probabilistic characterization of *lifetime*. To solve any problem in preventive maintenance we eventually will need information about the lifetime distribution function: the estimates of its parameters and/or a nonparametric estimate of the survival probability function.

By the very nature of *lifetime*, obtaining a complete sample of observations is practically impossible. In a laboratory test, we stop the experiment either at a prescribed time or after observing a prescribed number of failed items. Otherwise, the experiment becomes too costly and too time-consuming. Thus, for some items the lifetime is *censored*, i.e. our information about it has the form "the lifetime τ exceeds some value t_c."

Suppose we monitor a sample of 100 new type transmissions installed on new trucks. It is agreed that a failed transmission will be immediately reported, together with the corresponding mileage. But it may happen that the truck is damaged in an accident and/or the transmission has been dismantled. Then the information received might be the following: the lifetimes $t_1, t_2, \ldots, t_{45}$ of 45 failed transmissions have been observed and recorded. For the remaining 55 transmissions, the information has the following form: the lifetime τ_j exceeds some constant L_j, $j = 46, 47, \ldots, 100$.

An important case of incomplete information was described and formalized in the paper by Artamanovsky and Kordonsky (1970). After a fatigue failure is discovered in an operating aircraft, the whole fleet is grounded and inspected. The following information is recorded for aircraft i: the total time airborne T_i, and the presence or absence of a specific fatigue crack. So, for the ith aircraft, it is known only whether the fatigue life τ_i exceeds T_i or not.

This section describes a very useful nonparametric estimation procedure for the survival probability function $R(t)$, i.e. for the probability of survival past time t. This procedure does not assume any parametric framework, i.e. no a priori assumptions are made about the functional form of $R(t)$.

It will be assumed that the following information is available. A group of n identical items starts operating at $t_0 = 0$. During the operation, the items may fail and also may be withdrawn (lost) from the follow-up. It is assumed that the failure times are recorded, together with the number of items "alive" just prior to the failure time.

Suppose, failures occur at the instants $0 < t_1 < t_2 < \ldots < t_k$. Let n_j be the number of items "at risk" (i.e. "alive") just prior to t_j. Denote by w_j the number of items withdrawn from the observation in the interval between the $(j-1)$th and jth failure, i.e. in the interval (t_{j-1}, t_j), $t_0 = 0$. Obviously, $n_1 = n - w_1$, $n_2 = n_1 - 1 - w_1$, etc.

Kaplan and Meier (1958) suggested their famous nonparametric estimator $\hat{R}(t)$ of the survival probability $R(t)$, which has the following form:

$$\hat{R}(t) = \prod_{\{i: t_i \le t\}} (1 - 1/n_i) . \tag{3.1.1}$$

By (3.1.1), $\hat{R}(t)$ is right-continuous. It equals 1 for $0 \le t < t_1$; if observation (follow-up) stops at the instant of the kth failure t_k, while some items remain alive, $\hat{R}(t)$ is defined only for the interval $[0, t_k]$.

A heuristic derivation of the estimator (3.1.1) can be made as follows. Suppose our observations of the sample are carried out at the time instants t_i^*, $i = 1, 2, \ldots$. At t_i^* we record the number n_j^* of items alive just prior to the observation instant. In addition, assume that the number of failures ("deaths") d_j^* is known for the interval $\Delta_j = (t_{j-1}^*, t_j)$. Therefore, we know also the number of items w_j "lost" or withdrawn in this interval. Let τ be the item lifetime. Write $P(\tau > t_j^*) = R(t_j^*) = P(\text{ survival past } t_j^*)$. Obviously,

$$R(t_j^*) = P(\tau > t_0^*) P(\tau > t_1^* | \tau > t_0^*) \cdots P(\tau > t_j^* | \tau > t_{j-1}^*). \tag{3.1.2}$$

Table 3.1: ABS test data

failure number i	mileage t_i	n_i	$(n_i - 1)/n_i$	$\hat{R}(t_i)$
1	3 220	50	0.980	0.980
2	6 250	49	0.980	0.960
3	12 660	46	0.978	0.939
4	15 610	42	0.976	0.916
5	22 980	39	0.974	0.892
6	27 570	35	0.971	0.866
7	30 800	34	0.971	0.841
8	33 460	30	0.967	0.813
9	38 500	27	0.963	0.783
10	41 290	25	0.960	0.752
11	44 870	20	0.950	0.714
12	50 070	16	0.938	0.670

Let $p_i = P(\tau > t_i^* | \tau > t_{i-1}^*)$. A natural estimate of p_i is the quantity $\hat{p}_i = 1 - d_i^*/\hat{n}_i^*$, where $\hat{n}_i^*$ is the average number of items operational during the interval Δ_i. It seems reasonable to put $\hat{n}_j^* = n_j^* - w_j/2$. Thus the so-called *life-table estimate* $\bar{R}(t_j)$ of $R(t_j)$ is obtained as

$$\bar{R}(t_j) = \prod_{i=1}^{j} \hat{p}_i. \tag{3.1.3}$$

The Kaplan–Meier estimator can be viewed as the limiting case of the estimator (3.1.3). Imagine that the lengths of Δ_j tend to zero, and the number of these intervals tends to infinity. Some of these intervals which do not contain failures will contribute to (3.1.1) by a factor of 1. The only nontrivial contribution will come from those intervals which contain failures. For them, assuming that failure instants are separated from each other, $\hat{p}_i^* = 1 - 1/n_i^* = 1 - 1/n_i$. Thus we arrive at the Kaplan–Meier estimator also termed the product-limit (PL) estimator.

Example 3.1.1. Kaplan–Meier estimates for ABS units
A new experimentally designed automatic braking system (ABS) was installed in 50 cars. At ABS failures, the mileage of each car was recorded. In addition, some cars were removed from testing due to the failures of other parts or because of accidents. The experiment was designed in such a way that at the instant of the ABS failure or car removal, the corresponding mileage was recorded. The data and the computation of $\hat{R}(t)$ are summarized in Table 3.1.

It can be shown that under a rather general random withdrawal mechanism, $\hat{R}(t)$ is an unbiased estimator of $R(t)$. The following formula known in the literature as the Geenwood's formula gives an estimate of $Var[\hat{R}(t)]$:

$$\widehat{Var}[\hat{R}(t)] = [\hat{R}(t)]^2 \sum_{\{i:t_i \le t\}} (n_i(n_i-1))^{-1} . \tag{3.1.4}$$

An approximate $1-2\alpha$ confidence interval on the survival probability for a fixed value t can be obtained as

$$[\hat{R}(t) - z_\alpha(\widehat{Var}[\hat{R}(t)])^{0.5},\ \hat{R}(t) + z_\alpha(\widehat{Var}[\hat{R}(t)])^{0.5}], \tag{3.1.5}$$

where z_α is the α quantile of the standard $N(0,1)$ normal distribution.

3.2 Probability Paper

In this section we describe a popular graphical technique for analyzing lifetime data which uses so-called ***probability paper***. This technique is usually the first step in lifetime data analysis. By means of probability paper it is possible to check visually how closely the data follow the hypothetical distribution function. Furthemore, it provides a quick estimation of the distribution parameters. Very convenient is also the fact that the probability paper is applicable to a right censored samples, a feature which is very important for reliability applications.

Use of probability paper is restricted to location–scale families; see Definition 2.3.1.

Suppose, therefore, that the c.d.f $F(\cdot)$ of lifetime τ is such that

$$P(\tau \le t) = F\Big(\frac{t-a}{b}\Big),\ \ b > 0. \tag{3.2.1}$$

We assume that $F(\cdot)$ is a continuous and strictly increasing function. Then, for any p between 0 and 1, the equation

$$p = F\Big(\frac{t-a}{b}\Big) \tag{3.2.2}$$

has a single root denoted by t_p.

Definition 3.2.1: *p quantile*
The root $t = t_p$ of (3.2.2) is called the ***quantile*** of level p, or the p quantile.

The p quantile has a simple probabilistic meaning: the p part of the probability lies to the left of t_p:

$$P(\tau \le t_p) = p. \tag{3.2.3}$$

The median is the 0.5 quantile, and the lower quartile is the 0.25 quantile. Denoting by $F^{-1}(\cdot)$ the inverse function of $F(\cdot)$, we obtain that

$$\frac{t_p - a}{b} = F^{-1}(p), \tag{3.2.4}$$

or

$$t_p = a + bF^{-1}(p). \tag{3.2.5}$$

This equation shows that for a location–scale family there is a linear relationship between $F^{-1}(p)$ and the t_p. This is the basis for constructing the probability paper.

Let $t_{(1)}, t_{(2)}, \ldots, t_{(n)}$ be the *ordered* observations from the population $\tau \sim F((t-a)/b)$.

The *sample* counterpart of the c.d.f. is called the *empirical* c.d.f. and denoted by $\hat{F}_n(t)$. It is a stepwise function that jumps by $1/n$ at each $t_{(i)}$.

Definition 3.2.2: *The empirical distribution function* $\hat{F}_n(t)$
$\hat{F}_n(t) = 0$ for $t < t_{(1)}$ and $\hat{F}_n(t) = 1$ for $t \geq t_{(n)}$, and

$$\hat{F}_n(t) = k/n, \quad \text{for } t_{(k)} \leq t < t_{(k+1)}, \quad k = 1, \ldots, n-1\,. \tag{3.2.6}$$

Let t_p be the p quantile of $F(\cdot)$. Then it is easy to prove that $\hat{F}_n(t_p) \to p$ as $n \to \infty$ (since p is the probability that an observed lifetime is at most t_p). Then $t_{(k)}$ is the empirical value of the $p = k/n$ quantile, and we can write

$$\hat{p}_k = \frac{k}{n} \approx F\Big(\frac{t_{(k)} - a}{b}\Big), \tag{3.2.7}$$

or

$$t_{(k)} \approx a + b \cdot F^{-1}(\hat{p}_k)\,. \tag{3.2.8}$$

This relationship is the key for probability plotting: the observed lifetimes $t_{(k)}$ must be plotted against $F^{-1}(\hat{p}_k)$. $\hat{F}(t)$ changes from $(k-1)/n$ to k/n at $t_{(k)}$; plotting $t_{(k)}$ against $F^{-1}((k-0.5)/n)$ is recommended; (see e.g. Vardeman 1994, pp. 77, 78).

Then a straight line must be drawn by eye through the plotted points. Closeness of the points to this line is confirmation that the sample belongs to the population with the hypothetical c.d.f. $F(t)$.

To facilitate the use of probability paper, we present in Appendix D the Weibull and the normal probability paper, together with *Mathematica* code for producing the paper and plotting on it.

Let $\tau = \psi(X)$, ψ be a monotonically increasing function, and let the c.d.f. of τ belong to a location-scale family. Then

$$P(X \leq y) = P(\psi(X) \leq \psi(y)) = F\Big(\frac{\psi(y) - a}{b}\Big). \tag{3.2.9}$$

Therefore, changing the time scale from t to $\psi(t)$ enables us to obtain a probability plot for an r.v. X. We give two very important examples.

(i) Let $\tau = \log X \sim N(\mu, \sigma^2)$, i.e. X is lognormal. We write this as $X \sim logN(\mu, \sigma^2)$. Here $\psi(X) = \log X$. Thus, probability paper for the lognormal distribution is obtained from normal paper by replacing the time axis by a logarithmic time scale.

(ii) Let $\tau \sim \text{Extr}(a, b)$, i.e. $P(\tau > t) = 1 - \exp(-e^{(t-a)/b})$. Then $X = e^{\tau} \sim W(\lambda, \beta)$, with $\lambda = e^{-a}, \beta = b^{-1}$. Therefore, Weibull paper is obtained from the probability paper for the extreme-value family by a logarithmic change of time scale.

Example 3.2.1

Seven braking units were installed on experimental trucks. Five failures were observed in the experiment, at 2.25, 6.7, 37.6, 85.4 and 110.0 (in thousands of miles) . It was decided to stop the test after observing five failures. Two braking units survived 110 000 miles. The following table presents the data needed for a probability plot:

i	$t_{(i)}$	$(i - 0.5)/7$
1	2.25	0.071
2	6.7	0.214
3	37.6	0.357
4	85.4	0.50
5	110.0	0.642

Figure 3.1 shows 5 points with coordinates $(t_{(i)}, F^{-1}((i - 0.5)/7))$ plotted on Weibull paper. They follow quite closely a straight line. We presume that if the two remaining points had been observed, they would have been close to the line drawn. We have, therefore, some evidence that the underlying c.d.f. is Weibull. It is important to note that we are dealing with an incomplete sample! Note also that in most applications, the ***left*** tail of the distribution is of greatest importance. In practice, we will hardly ever be able to observe all order statistics.

Weibull paper also enables us to obtain parameter estimates. The estimation of λ uses a special line which corresponds to the probability $1 - e^{-1} \approx 0.632$. The time value which corresponds to the intersection of the line drawn and this special line is the estimator of $1/\lambda$. In our case, $\hat{\lambda} = 1/115$.

To estimate the Weibull shape parameter β, we suggest the following procedure. Read from the graph of the line drawn the time value t_p which corresponds to a certain p-value, say $p = 0.35$. We see that $t_{0.35} \approx 27$. Then use the following formula:

$$\hat{\beta} = \frac{\log(-\log(1-p))}{\log(t_p \hat{\lambda})}. \tag{3.2.10}$$

For our data we obtain $\hat{\beta} = 0.58$.

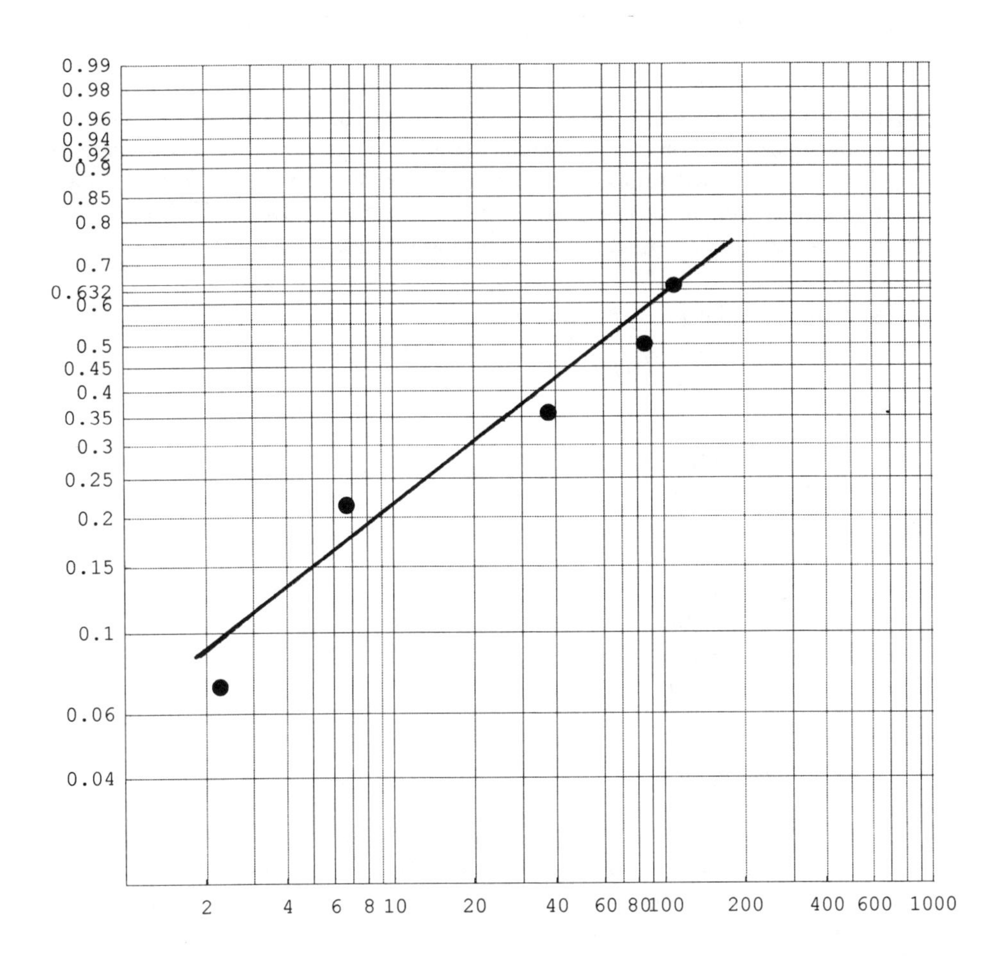

Figure 3.1. The Weibull plot for Example 3.2.1

To facilitate the use of Weibull paper, we present in Appendix D *Mathematica* code for plotting on it. To activate this program, define

(i) the observed lifetimes as "tw", see *In*[*1*] and

(ii) "nobsw", the total number of observations.

The program in Appendix D is designed for the range of observations [1, 1 000]. If the observations fall outside this range, rescale the range. For example, if the maximal observation is 100 000 cycles, express the results in hundreds of cycles.

There is an ample literature on using graphical methods for data analysis and estimation. We refer the interested reader to Abernethy et al (1983), Dodson (1994) and Vardeman (1994).

Example 3.2.2: Fatigue test data for cyclic bending of a cantilever beam
The following are the results of a fatigue experiment. A cantilever beam made of alloy B-95 was bent with maximal cyclic stress 30 kg mm^{-2} (Kordonsky 1966). Of 22 specimen put on test, 13 failed in the interval $[0, 1.25 \times 10^5]$, see Table 3.2. $N(i)$ is the lifetime in the units of 100 000 cycles.

Table 3.2: The fatigue test results

i	$N(i)$	$\log_{10} N(i)$
1	0.53	4.724
2	0.65	4.813
3	0.76	4.881
4	0.80	4.905
5	0.87	4.940
6	0.90	4.954
7	0.902	4.955
8	1.02	5.009
9	1.07	5.029
10	1.074	5.031
11	1.09	5.037
12	1.16	5.064
13	1.22	5.085

Figure 3.2 is the plot on normal paper of the lifetimes of the noncensored observations obtained by means of the *Mathematica* program presented in Appendix D. In our case, we have to supply the program with the following data:

(i) the logarithms of the lifetimes put into the list "tn", the third column of Table 3.2;

(ii) the total number of observations (including censored), "nobsn." In our case, nobsn=22.

(iii) the first argument of PlotRange should include the minimal and the maximal observed lifetime (in logarithms). We took $[4.6, 5.2]$; see Fig. 3.2.

Figure 3.2 shows that the dots closely follow a straight line. Assuming that the censored lifetimes belong to the same population, it seems very plausible that the actual lifetimes follow the lognormal distribution. (Recall that if $\log \tau \sim N(\mu, \sigma)$, then $\tau \sim logN(\mu, \sigma)$).

Parameter estimation from the normal plot is very simple. Draw a straight line L through the plotted points. The intersection of L with the horizontal line marked 0.5 is the estimate of μ. In our case, $\hat{\mu} = 5.060$. To obtain the estimate of σ, read the abscissa x_0 of the intersection of L with the horizontal line just above the line marked 0.15 (this line corresponds to the probability 0.159): $x_0 = 4.893$. The estimate of σ is $\hat{\sigma} = \hat{\mu} - x_0 = 5.060 - 4.893 = 0.167$.

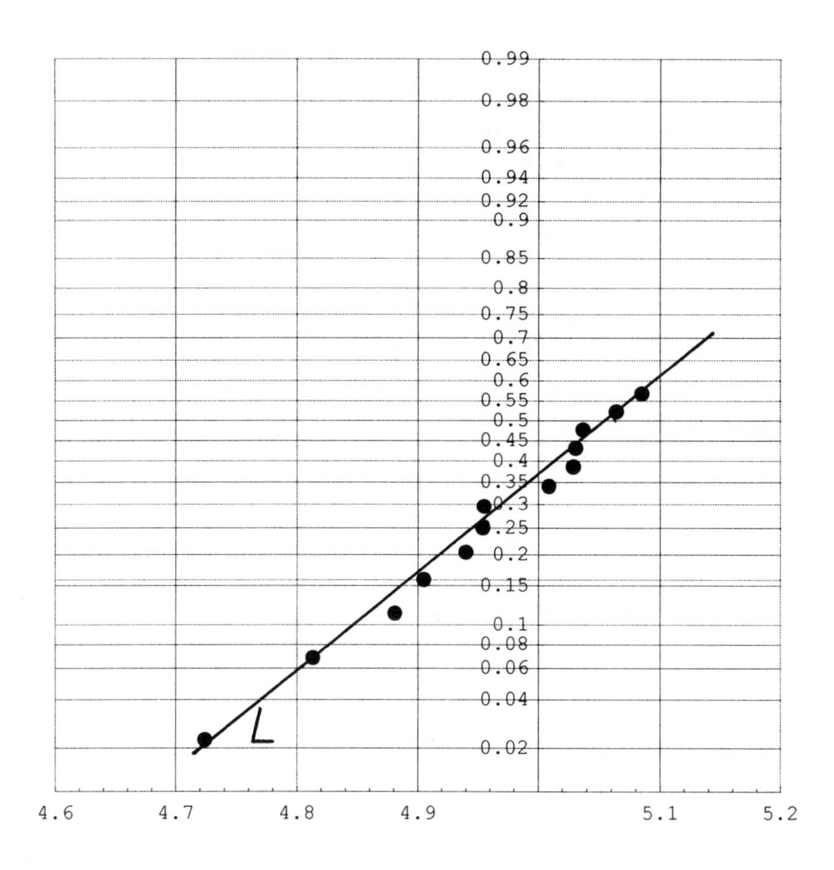

Figure 3.2. The normal plot for Example 3.2.2

Remark

So far we have described the use of probability paper for complete or right-censored samples. Could we draw a probability plot for data subject to more complicated censoring? A practical solution would be to use the Kaplan–Meier estimate $\hat{R}(t)$ to produce a substitute for the empirical distribution function. Put $\hat{F}(t) = 1 - \hat{R}(t)$. Suppose that at the point t_i, $\hat{F}(t)$ jumps from p_i^* to p_i^{**}. Then calculate $p_i = (p_i^* + p_i^{**})/2$ and plot $(t_i, F^{-1}(p_i))$ on the probability paper. In other words, use p_i instead of $(i - 0.5)/n$.

3.3 Parameter Estimation for Censored and Grouped Data

3.3.1 The Maximum Likelihood Method

Suppose we observe a sample $\{t_1, t_2, \ldots, t_n\}$ from a population with density function $f(t; \alpha, \beta)$. (For the sake of simplicity, we assume that $f(\cdot)$ depends on two unknown parameters α and β .) The standard maximum likelihood method of estimating the parameters works as follows.

(i) Write the ***likelihood function***

$$Lik = Lik(\alpha, \beta; t_1, \ldots, t_n) = \prod_{i=1}^{n} f(t_i; \alpha, \beta) \,. \tag{3.3.1}$$

(ii) Find the values of α and β (depending on the observed sample values) which ***maximize*** the likelihood function. These values $\hat{\alpha} = \hat{\alpha}(t_1, t_2, \ldots, t_n)$ and $\hat{\beta} = \hat{\beta}(t_1, t_2, \ldots, t_n)$ are called the ***maximum likelihood estimates*** (MLEs) of the parameters.

In practice, the MLE are found by taking the logarithm of the likelihood function

$$logLik = \sum_{i=1}^{n} \log f(t_i; \alpha, \beta), \tag{3.3.2}$$

and by solving the system of equations

$$\frac{\partial logLik}{\partial \alpha} = 0, \quad \frac{\partial logLik}{\partial \beta} = 0. \tag{3.3.3}$$

Assume that the solution $(\hat{\alpha}, \hat{\beta})$ of (3.3.3) does exist, is unique and corresponds to the maximum of the likelihood function. Then $(\hat{\alpha}, \hat{\beta})$ are the maximum likelihood estimates.

Typically, the MLEs have good statistical properties and in some cases coincide with the optimal estimates.

Remark

Finding the MLE as the stationary point via the system of equations (3.3.3) does ***not*** work in one important case: when the support of the density function $f(t; \cdot, \cdot)$ depends on the parameters. An example is the exponential density with a threshold parameter: $f(t; \lambda, a) = \lambda e^{-\lambda(t-a)}, \ t \geq a$. Here the ***support*** depends on the parameter a, and the maximum of the likelihood function is not attained at the point where partial derivatives are zero. But the principle that the MLE should maximize the likelihood function remains valid. We demonstrate later for the exponential distribution with threshold parameter that the MLE can be found directly from the expression for *logLik*; see Exercise 4 in Sect. 3.4.

Remark
An important property of the maximum likelihood method is the so-called ***invariance***. Suppose that we want to find the MLE $\hat{\psi}$ of some parametric function $\psi(\alpha, \beta)$. By definition, $\hat{\psi}$ is the value of the function $\psi(\alpha, \beta)$ which corresponds to the global maximum of the likelihood function. We may be tempted to express the likelihood function as a function of ψ and to maximize it. This, however, is somewhat tedious, and there is a much easier solution: the MLE of ψ is obtained via the formula

$$\hat{\psi} = \hat{\psi}(\hat{\alpha}, \hat{\beta}). \tag{3.3.4}$$

For example, the p quantile for a location-scale family $F((t-a)/b)$ equals $t_p = a + b \cdot F^{-1}(p)$. (Verify it!) Suppose that we have the MLEs $\hat{a}$ and $\hat{b}$ of a and b, respectively. Then the MLE of the p quantile t_p is

$$\hat{t}_p = \hat{a} + \hat{b}F^{-1}(p). \tag{3.3.5}$$

The main advantage of the maximum likelihood method is that it can be easily adjusted for ***incomplete*** data, such as censored and/or grouped samples. This we demonstrate in the next subsection.

3.3.2 Maximum Likelihood Function for Censored and Grouped Data

Censoring by an order statistic

The population has density function $f(t;\alpha,\beta)$ and c.d.f. $F(t;\alpha,\beta)$. The information available consists of the first k order statistics $t_{(1)}, t_{(2)}, \ldots, t_{(k)}$. The only information about the remaining $n-k$ observations is that they exceed $t_{(k)}$. This type of data (termed type II censoring) is typical in ***reliability testing***: n devices are subjected to a test which terminates, because of time or cost restrictions, exactly when the kth failure is observed. The likelihood function has the following form:

$$Lik = \prod_{i=1}^{k} f(t_{(i)};\alpha,\beta) \cdot [1 - F(t_{(k)};\alpha,\beta)]^{n-k}\ . \tag{3.3.6}$$

Example 3.3.1: Exponential distribution
Let $f(t,\lambda) = \lambda e^{-\lambda t}$. Then

$$logLik = k \log k - \lambda(t_{(1)} + \ldots + t_{(k)} + (n-k)t_{(k)}). \tag{3.3.7}$$

The expression in brackets in (3.3.7) is the total time on test, TTT. The equation $\partial logLik/\partial\lambda = 0$ gives

$$\hat{\lambda} = k/\text{TTT}. \tag{3.3.8}$$

It is worth recalling that TTT $\sim Gamma(k, \lambda)$; see Exercise 4, Sect. 2.4. This fact is useful in investigating the statistical properties of $\hat{\lambda}$.

Censoring by a constant

The lifetime data are censored by a fixed constant T (so-called type I censoring). A typical example is testing n devices during a fixed period of time $[0, T]$. Denote the observed lifetimes $t_{(1)}, t_{(2)}, \ldots, t_{(k)}$, exactly as in the previous case, with the difference being that here k is *random* and is not fixed in advance. To make the maximum likelihood method work, we must observe at least one noncensored lifetime. The likelihood function is similar to type II censoring:

$$Lik = \prod_{i=1}^{k} f(t_{(i)}; \alpha, \beta) \cdot [1 - F(T; \alpha, \beta)]^{n-k}. \tag{3.3.9}$$

Random noninformative censoring

Suppose that the lifetime τ_i of the ith item can be observed only during a random time interval $[0, T_i]$. The r.v. $T_i \sim G(t)$ is assumed to be independent of τ_i. For example, the lifetime of a transmission can be observed on the time interval $[0, T_i]$, where T_i is the actual mileage done by the ith truck during the warranty period, say one year. If the transmission fails during the first year, we will know the actual value of $\tau_i = t_i$ *and* also that $T_i > t_i$. Otherwise, we observe the truck one-year mileage $T_i = x_i$ and know only that τ_i *exceeds* x_i. We assume that $\tau_i \sim f(t; \alpha, \beta)$, and that the c.d.f. of τ_i is $F(t; \alpha, \beta)$.

The following assumption is important: the c.d.f. of T_i does *not depend* on the parameters (α, β), which justifies the name *noninformative* censoring.

To write the likelihood function we need a notation $g(t)$ for the density function and the c.d.f. of T_i. It is convenient to introduce an indicator random variable δ_i which equals 1 if $\tau_i \le T_i$, and 0 otherwise. Denote by y_i the observed lifetime or the censoring time for the ith item. Then the contribution of the ith item to the likelihood function is

$$Lik_i = [f(y_i; \alpha, \beta) \cdot (1 - G(y_i)]^{\delta_i} \cdot [g(y_i) \cdot (1 - F(y_i; \alpha, \beta))]^{(1-\delta_i)}. \tag{3.3.10}$$

Now the log-likelihood function becomes $logLik = \prod_{i=1}^{n} Lik_i$:

$$logLik = \sum_{i=1}^{n} \log \left[f(y_i; \alpha, \beta)^{\delta_i} \cdot (1 - F(y_i; \alpha, \beta))^{(1-\delta_i)} \right] + H(y_1, \ldots, y_n). \tag{3.3.11}$$

Here the last term $H(\cdot)$ depends only on $g(y_i)$ and $G(y_i)$ and does *not* contain unknown parameters. Since we proceed by taking partial derivatives of the log-likelihood with respect to α and β, this H-term can simply be omitted.

Quantal response data

We already mentioned the work of Artamanovsky and Kordonsky (1970), in which the following important situation was studied. For the ith item, the information about its lifetime is the following: either the item failed during the interval $[0, y_i]$, where y_i is a known quantity, or survived past y_i. The exact time of the failure is not known. This type of yes/no data about the failure time is called *quantal*. It is typical for periodic inspections and follow-up studies.

For the sake of simplicity, let us number the "yes" responses (i.e. the failed items) using the index i ranging from 1 to r. The nonfailed items have numbers from $r+1$ to n (the "no" response). Then the likelihood function is

$$Lik = \prod_{i=1}^{r} F(y_i; \alpha, \beta) \cdot \prod_{i=r+1}^{n} (1 - F(y_i; \alpha, \beta)) \ . \tag{3.3.12}$$

It has been proved in the above cited paper that for a location-scale family, the system of likelihood equations for this case has a unique solution if at least one "yes" and one "no" response have been observed.

Example 3.3.2: Testing n items on the interval $[0, y]$
Consider $F(t; \lambda) = 1 - e^{-\lambda t}$, all $y_i = y$. r items have failed on $[0, y]$ and $n - r$ have survived past y. Here $Lik = (1 - e^{-\lambda y})^r e^{-\lambda y(n-r)}$, so

$$logLik = r \log(1 - e^{-\lambda y} - \lambda y(n - r).$$

Hence

$$\frac{\partial logLik}{\partial \lambda} = rye^{-\lambda y}/(1 - e^{-\lambda y}) - y(n - r) = 0,$$

which gives

$$\hat{\lambda} = y^{-1} \log \frac{n}{(n - r)}.$$

Grouped data

Another situation often arising in lifetime testing and/or in follow-up studies is the following: the time axis is divided into nonoverlapping time intervals $I_1 = (t_0, t_1], I_2 = (t_1, t_2], \dots, I_k = (t_{k-1}, t_k], I_{k+1} = (t_k, t_{k+1})$, where $t_0 = 0$, $t_{k+1} = \infty$. N items are put on test (or are followed up) at $t = 0$, and the information available is that N_j of them have failed in the interval $I_j, j = 1, \dots, k+1$,

$\sum_{j=1}^{k+1} N_j = N$. One of the first monographs devoted to parameter estimation for grouped data was by Kulldorf (1961).

Assume that the items fail independently and that their lifetime has c.d.f. $F(t;\alpha,\beta)$. The likelihood function is

$$Lik = \prod_{i=1}^{k+1} \Big[F(t_i;\alpha,\beta) - F(t_{i-1};\alpha,\beta)\Big]^{N_i}. \tag{3.3.13}$$

Example 3.3.3: Fatigue test data
Twenty-five specimens made from duralumin alloy were tested over a period of 90 million cycles. The following data were recorded: 7 units failed on the interval $[0,45]$, 13 units failed on the interval $(45,90]$, 5 units survived the test. Previous experience says that the lifetime follows the Weibull distribution, $F(t;\lambda,\beta) = 1-\exp[-(\lambda t)^\beta]$. Prior information suggests that the scale parameter λ is in the range 0.008–0.02, and that the shape parameter β lies between 2 and 4. The likelihood function has the following expression:

$$\begin{aligned} Lik &= \Big(1-\exp[-(\lambda\cdot 45)^\beta]\Big)^7 \\ &\quad\times\Big(\exp[-(\lambda\cdot 45)^\beta] - \exp[-(\lambda\cdot 90)^\beta]\Big)^{13}\exp[-5(\lambda\cdot 90)^\beta]\,. \end{aligned}$$

Let us show how to use *Mathematica* to find the MLEs $\hat{\lambda}$ and $\hat{\beta}$.

In the printout presented as Fig. 3.3, *In[1]* defines the log-likelihood function as logLik, and its partial derivatives f1 and f2 with respect to λ and β, respectively. The operator "FindRoot" solves the system (3.3.3), and *Out*[*4*] is its solution: $\lambda = 0.0137$ and $\beta = 2.29$. This operator requires the initial values for the variables to be defined. As such were taken $\lambda = 0.014$ and $\beta = 3$, near the middle of the the prior range for these parameters. Since the point where both partial derivatives are zero is not necessarily the maximum point, it is highly advisory to make a contour plot of the log-likelihood function. This is carried out by means of the operator "ContourPlot"; see *In*[*5*]. The plot provided by *Out*[*5*] leaves no doubt that $\hat{\lambda},\hat{\beta}$ are indeed the MLEs.

3.3.3 Finding Maximum Likelihood Estimates for a Censored Sample: Weibull and lognormal Distribution

Weibull distribution

In this subsection we denote by $t_1, t_2, \ldots, t_r$ the *noncensored* observations, and by $t_{r+1}, \ldots, t_n$ the censored ones. For example, for type II censoring, $t_{r+1}, \ldots, t_n$ are equal to the largest observed order statistic.

It is more convenient to work not with the actual observed (or censored) lifetimes but rather with their logarithms. So let

$$y_i = \log t_i,\ i = 1,\ldots,r,r+1,\ldots,n. \tag{3.3.14}$$

```
In[1]:= logLik = 7 * Log[1 - Exp[- (45 λ) ^β]] +
           13 * Log[Exp[- (45 λ) ^β] - Exp[- (90 λ) ^β]] - 5 * (90 λ) ^β;
        f1 = D[logLik, λ];
        f2 = D[logLik, β];
        FindRoot[{f1 == 0, f2 == 0}, {λ, 0.014}, {β, 3}]

Out[4]= {λ → 0.0136744, β → 2.29257}

In[5]:= ContourPlot[logLik, {λ, 0.01, 0.016}, {β, 2, 3}, Contours → 30]
```

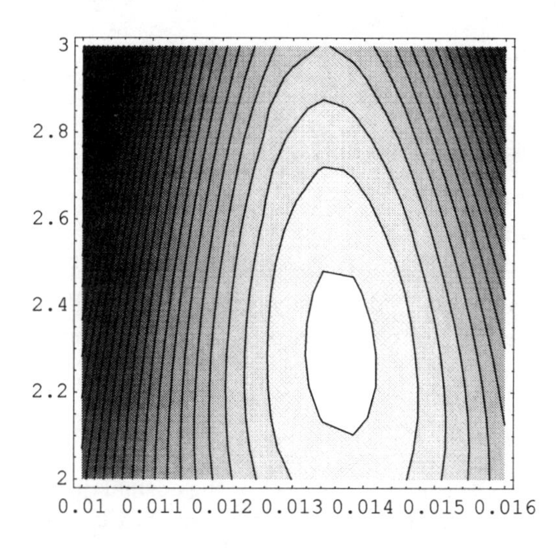


```
Out[5]= - ContourGraphics -
```

Figure 3.3. *Mathematica* printout for solving Example 3.3.3

Recall that if the original observations t_i come from the Weibull distribution, then their logarithms have the extreme-value distribution Extr(a, b), where a and b are expressed through the Weibull parameters:

$$\lambda = e^{-a}, \ \beta = 1/b\,. \tag{3.3.15}$$

The system of likelihood equations has the following form:

$$\partial logLik/\partial a = -r/b + b^{-1}\sum_{i=1}^{n} e^{(y_i - a)/b} = 0,$$

$$\partial logLik/\partial b = -r/b - b^{-2}\sum_{i=1}^{r}(y_i - a) + b^{-2}\sum_{i=1}^{n} e^{(y_i-a)/b}(y_i - a) = 0\,.$$

From the computational point of view, it is always easier to solve one nonlinear equation than a system of two. The first of the above equations allows a to be expressed as a function of b:

$$a = b\log\left(r^{-1}\sum_{i=1}^{n} e^{y_i/b}\right). \tag{3.3.16}$$

Using (3.3.16), we obtain an equation for b only:

$$\frac{\sum_{i=1}^{n} y_i e^{y_i/b}}{\sum_{i=1}^{n} e^{y_i/b}} = r^{-1}\sum_{i=1}^{r} y_i + b\,. \tag{3.3.17}$$

This equation usually has a single root and is easily solved numerically. Denote its solution by $\hat{b}$. This will be the desired MLE of b. Now substitute it into (3.3.16) and obtain $\hat{a}$.

Lognormal distribution

We know that if the logarithms of actual lifetimes follow the normal distribution, then the lifetimes themselves are lognormal. So, we process the logarithms (y_i, $i = 1, \ldots, n$) as if they are normally distributed. The procedure for finding the MLEs for the normal case with censored observations is based on a rather efficient iterative procedure, which is a version of the so-called Expectation-Maximization algorithm; see e.g. Lawless (1982, Chap. 5). To describe it, we need some notation.

Let

$$\phi(x) = (\sqrt{2\pi})^{-1}e^{-x^2/2}, \ \ R(x) = \int_x^{\infty} \phi(t)dt\,. \tag{3.3.18}$$

Define also

$$U(x) = \phi(x)/R(x); \ \ \gamma(x) = U(x)\cdot(U(x) - x)\,. \tag{3.3.19}$$

Denote by μ and σ the parameters of the normal distribution. Define $x_i = (y_i - \mu)/\sigma$, $w_i = y_i$ for $i = 1, \ldots, r$, i.e. for the noncensored observations, and $w_i = \mu + \sigma \cdot U(x_i)$ for $i = r+1, \ldots, n$, i.e. for the censored observations.

Now use the following iterative procedure.

Step 0. Define initial estimates of μ and σ. Denote them as μ_0 and σ_0.

Step 1. Use μ_0 and σ_0 to compute w_i from the above formulas, for $i = 1, 2, \ldots, n$.

Step 2. Compute μ_1 and σ_1 from the equations

$$\mu_1 = \sum_{i=1}^{n} w_i/n \tag{3.3.20}$$

and

$$\sigma_1^2 = \frac{\sum_{i=1}^{n}(w_i - \mu_0)^2}{r + \sum_{i=r+1}^{n} \gamma(x_i)}. \tag{3.3.21}$$

Go to step 1 by setting $\mu_0 := \mu_1$ and $\sigma_0 := \sigma_1$.

Repeat steps 1,2 until procedure converges.

It is easy to implement the above procedure on *Mathematica* because it has built-in operators for computing $R(x)$.

We have considered in this section the most useful parameter estimation techniques for incomplete data. This type of statistical inference is probably the most relevant to the implementation of reliability theory and preventive maintenance models to real-life situations.

There is an ample literature on this topic. The reader can find a lot of information in Lawless (1982) and Gertsbakh (1989). These sources also give numerous references on inference from censored, grouped and quantal-type data.

3.3.4 Point and Confidence Estimation of Location and Scale Parameters Based on Linear Combination of Order Statistics

The purpose of this subsection is to describe methods of parameter estimation for incomplete data when the sample is drawn from a location–scale family. The parameter estimates are linear combinations of observed order statistics. These methods are easy to use and the quality of the corresponding estimators is comparable to the quality of maximum likelihood estimators.

Suppose that a random sample $X_1, X_2, \ldots, X_n$ is drawn from a population of location–scale type; see Definition 2.3.1. Recall that $X_i \sim F_0((t-\mu)/\sigma)$. Denote by $X_{(i)}$ the ith order statistic, and let $Z_i = (X_i - \mu)/\sigma$. Obviously Z_i is parameter-free, $P(Z_i \leq t) = F_0(t)$. Then one can write the following set of equations:

$$X_{(i)} = \mu + \sigma m_{(i)} + \sigma(Z_{(i)} - m_{(i)}), \tag{3.3.22}$$

where $m_{(i)} = E[Z_{(i)}]$. The index i in (3.3.22) takes on the ordinal numbers of the observed order statistics. For example, we observe the second, third and fifth ordered observations. Then $i = 2, 3$ and 5. To simplify the exposition, we assume that we observe the first r order statistics, i.e. we are in a situation of type II censoring. Then in (3.3.22), $i = 1, 2, \ldots, r$.

Since Z_i is parameter-free, the values of m_i can viewed as known. Thus the set of r equations (3.3.22) contains two unknowns, μ and σ; the term $\epsilon_i = Z_{(i)} - m_{(i)}$ is a zero-mean random variable. Expression (3.3.22) has matrix form

$$\mathbf{X} = \mathbf{D}\beta + \sigma\epsilon, \tag{3.3.23}$$

where $\mathbf{X}' = (X_{(1)}, \ldots, X_{(r)})$ (the prime denotes transposition); β is a column vector with components μ, σ; ϵ is a column vector with components $\epsilon_1, \ldots, \epsilon_r$. $\mathbf{D}$ is the matrix

$$\mathbf{D} = \begin{bmatrix} 1 & m_{(1)} \\ 1 & m_{(2)} \\ \ldots & \ldots \\ \ldots & \ldots \\ 1 & m_{(r)} \end{bmatrix}.$$

Expression (3.3.23) is a standard *linear regression* relationship. To make the regression mechanism work, we have to compute the variance–covariance matrix

$$\mathbf{V} = ||Cov[\epsilon_i, \epsilon_j]||, \quad i, j = 1, 2, \ldots, r\,. \tag{3.3.24}$$

This can be done since the ϵ_i are parameter free. It is known from regression theory (see e.g. Seber 1977, Chap. 3) that:

(i) the minimal variance linear unbiased estimators of μ and σ are

$$(\hat{\mu}, \hat{\sigma})' = (\mathbf{D}'\mathbf{V}^{-1}\mathbf{D})^{-1}\mathbf{D}'\mathbf{V}^{-1}\mathbf{X}; \tag{3.3.25}$$

(ii) the covariance matrix of $(\hat{\mu}, \hat{\sigma})$ is

$$\sigma^2(\mathbf{D}'\mathbf{V}^{-1}\mathbf{D})^{-1}\,. \tag{3.3.26}$$

$\hat{\mu}, \hat{\sigma}$ are called the ***best linear unbiased estimators*** (BLUEs). As follows from (3.3.25), these estimators have the following general form:

$$\hat{\mu} = \sum_{i=1}^{r} a(n, r; i) X_{(i)},$$

$$\hat{\sigma} = \sum_{i=1}^{r} b(n, r; i) X_{(i)}.$$

Recall that the regression method works for any collection of order statistics. If we have at our disposal the order statistics with indices $i_1, i_2, \ldots, i_r$, the same method applies with obvious changes in the $\mathbf{D}$ matrix and in the $\mathbf{V}$ matrix.

The normal case has been given extensive theoretical and numerical consideration. If we have a sample which, by assumption, follows the lognormal distribution, we have to process the *logarithms* of the observations as a sample from the normal distribution. The paper by Sarhan and Greenberg (1962) contains tables of the coefficients $a(n,r;i), b(n,r;i)$, for a wide range of n and r values. It considers type II censored samples, as well as left- and right-censored samples. We recommend the use of the above source for the normal case.

If we have a sample from the Weibull distribution then the logarithms of the observations follow the so-called extreme-value distribution (2.3.15). BLUE estimators for this case have been investigated by Lieblein and Zelen (1956).

To facilitate the numerical work in calculating the BLUEs, we present in Appendix B the **V** matrices for the normal and the extreme-value case, for $n = 8, 10$ and $n = 15$, together with the mean values of the order statistics $m_{(i)}$.

An interesting theoretical and practical issue is to compare the maximum likelihood estimators and the BLUEs. A suitable criterion for comparison is the mean square error (m.s.e.). Recall that the m.s.e. of an estimator $\hat{\theta}$ is defined as

$$m.s.e.[\hat{\theta}] = Var[\hat{\theta}] + E[(\hat{\theta} - \theta)^2]. \tag{3.3.27}$$

For the extreme-value distribution, both methods provide close results in the absence of of heavy censoring. If heavy censoring is present, the MLEs should be used in preference to the BLUEs.

So far we have demonstrated several methods of obtaining point estimators for location and scale parameters. Now let us discuss confidence estimation for these parameters. Let $\hat{\mu}$ and $\hat{\sigma}$ be the estimators of the location and scale parameters, respectively. These estimators are functions of the observed order statistics $X_{(1)}, \ldots, X_{(r)}$:

$$\hat{\mu} = \psi_1(X_{(1)}, \ldots, X_{(r)}), \tag{3.3.28}$$

$$\hat{\sigma} = \psi_2(X_{(1)}, \ldots, X_{(r)}) \,. \tag{3.3.29}$$

Definition 3.3.1: *Equivariance*
If, for any d and any $c > 0$, the estimators ψ_1, ψ_2 satisfy the following properties:

$$\psi_1(cX_{(1)} + d, \ldots, cX_{(r)} + d) = c\psi_1(X_{(1)}, \ldots, X_{(r)}) + d, \tag{3.3.30}$$
$$\psi_2(cX_{(1)} + d, \ldots, cX_{(r)} + d) = c\psi_2(X_{(1)}, \ldots, X_{(r)}), \tag{3.3.31}$$

then they are termed *equivariant.*

Expression (3.3.30) means that if all observations X_i undergo the transformation $X_i^* = cX_i + d$, then the estimator of the *location* parameter undergoes the same transformation. Expression (3.3.31) means that for the same transformation, the estimator of the *scale* parameter is multiplied by factor c and remains insensitive to d.

We state without proof that the BLUEs of μ and σ are equivariant.

The following notion plays a central role in confidence estimation.

Definition 3.3.2: *Pivotal quantity*
Let the c.d.f. of X_i depend on some parameter θ. Then a function of $X_1, \ldots, X_n$ and θ, $Q(X_1, X_2, \ldots X_n; \theta)$, is called a ***pivotal quantity*** if the distribution of Q is parameter-free.

To show how a pivotal quantity can be used for confidence estimation, let us recall the famous expression

$$q = \frac{\sqrt{n-1}(\overline{X} - \mu)}{\left(\sum_{i=1}^{n}(X_i - \overline{X})^2/n\right)^{0.5}}, \tag{3.3.32}$$

where $X_i \sim N(\mu, \sigma)$, and $\overline{X} = \sum_{i=1}^{n} X_i/n$ is the sample average. The reader knows from the theory of statistics (see e.g. Devore 1982, p. 259) that q is distributed according to the so-called t distribution (with $n-1$ degrees of freedom), which does not depend on the parameters μ and σ. Then the confidence interval for μ is obtained from the following relationship:

$$P\big(t_\alpha \le q \le t_{1-\alpha}\big) = 1 - 2\alpha, \tag{3.3.33}$$

where $t_\alpha, t_{1-\alpha}$ are the corresponding quantiles of the t distribution. Transforming the inequalities inside $P(\cdot)$ in (3.3.33) leads to the following well-known $1 - 2\alpha$ confidence interval:

$$\left[\overline{X} - \frac{t_\alpha}{\sqrt{n}}s,\ \overline{X} + \frac{t_{1-\alpha}}{\sqrt{n}}s\right], \tag{3.3.34}$$

where $s = \left(\sum_{i=1}^{n}(X_i - \overline{X})^2/(n-1)\right)^{0.5}$.

Let $G_Q(t)$ be the c.d.f. of the pivotal quantity Q. Then a confidence interval for θ could be obtained by using the relationship

$$P\big(q_\alpha \le Q(X_1, \ldots X_n; \theta) \le q_{1-\alpha}\big) = 1 - 2\alpha, \tag{3.3.35}$$

where $q_\alpha, q_{1-\alpha}$ are the corresponding quantiles of $G_Q(t)$. The desired confidence interval on θ will be obtained if we succeed in "pivoting," i.e. in solving the inequalities inside $P(\cdot)$ in (3.3.35) with respect to θ. Let us show how this procedure works if we have at our disposal equivariant estimators satisfying (3.3.30) and (3.3.31).

Theorem 3.3.1
Let ψ_1 and ψ_2 satisfy (3.3.30), (3.3.31), and let t_p be the p quantile of the r.v. $Y \sim F_0((t-\mu)/\sigma)$. Then

$$A_1 = \frac{\psi_1 - \mu}{\psi_2},\ A_2 = \frac{\psi_2}{\sigma},\ A_3 = \frac{\psi_1 - \mu}{\sigma},\ A_4 = \frac{\psi_1 - t_p}{\psi_2} \tag{3.3.36}$$

are pivotal quantities.

Proof
Let $Y_i = (X_i - \mu)/\sigma$ be i.i.d., $Y_i \sim F_0(t)$. Then $Y_{(i)} = (X_{(i)} - \mu)/\sigma$, $i = 1, \ldots, n$. Obviously, $\psi_2(Y_{(1)}, \ldots, Y_{(r)})$ is a pivotal quantity. It equals

$$\psi_2((X_{(1)} - \mu)/\sigma, \ldots, (X_{(r)} - \mu)/\sigma)) = \psi_2(X_{(1)}, \ldots, X_{(r)})/\sigma = A_2.$$

Similarly, $\psi_1(Y_{(1)}, \ldots, Y_{(r)})$ is a pivotal quantity. By (3.3.30), it equals

$$(\psi_1(X_{(1)}, \ldots, X_{(r)}) - \mu)/\sigma = A_3.$$

A_1 is a pivotal quantity because $A_1 = A_3/A_2$. For a location–scale family, $t_p = \mu + \sigma t_p^0$, where $F_0(t_p^0) = p$. A_4 is a pivotal because

$$\begin{aligned} A_4 &= (\psi_1(X_{(1)}, \ldots, X_{(r)}) - t_p)/\psi_2(X_{(1)}, \ldots, X_{(r)}) \\ &= \frac{(\psi_1(\mu + \sigma X_{(1)}, \ldots, \mu + \sigma X_{(r)} - \mu - \sigma t_p^0)}{\psi_2(\mu + \sigma X_{(1)}, \ldots, \mu + \sigma X_{(r)})} \\ &= (\psi_1(Y_{(1)}, \ldots, Y_{(r)}) - t_p^0)/\psi_2(Y_{(1)}, \ldots, Y_{(r)}). \end{aligned}$$

If the quantiles of the c.d.f. of the pivotals are known, then the use of Theorem 3.3.1 for obtaining confidence intervals is straightforward. For example, let $A_4 \sim G_4(t)$, $G_4(t_\alpha) = \alpha$ and $G_4(t_{1-\alpha}) = 1 - \alpha$. Then

$$P\big(t_\alpha \le \frac{\psi_1 - t_p}{\psi_2} \le t_{1-\alpha}\big). \tag{3.3.37}$$

Solving the inequality inside $P(\cdot)$ with respect to t_p (assuming $\psi_2 > 0$) produces the following $1 - 2\alpha$ confidence interval on t_p:

$$[\psi_1 - \psi_2 t_{1-\alpha}, \psi_1 - \psi_2 t_\alpha]\ . \tag{3.3.38}$$

The only remaining obstacle is finding the quantiles of the pivotals. The case with the t distribution is a rare exception because the c.d.f. of the pivotal is available in a closed form. For practical purposes, the following Monte Carlo simulation procedure could be used to find the quantiles of the pivotals.

Step 1. Generate a random sample $Y_1, \ldots, Y_n$ from the distribution $F_0(t)$.
Step 2. Calculate the statistics ψ_1, ψ_2 and the pivotal quantities $A_1, \ldots, A_4$. Denote by $A_j^{(k)}$, $j = 1, 2, 3, 4$, the kth replica of the respective pivotal.

Do N (say, $N = 100\,000$) iterations of steps 1, 2. Order the samples of $A_j^{(k)}$, $k = 1, 2, \ldots, N$; $j = 1, 2, 3, 4$. For each j, take as an estimate of the corresponding α quantile the $[\lfloor N\alpha \rfloor]$th ordered observation.

For example, for $N = 100\,000$ and $\alpha = 0.025$, $t_{0.025} = A_j^{(2500)}$, i.e. the quantile estimate is the 2500th ordered observation. Similarly, $t_{0.975} = A_j^{(97\,500)}$.

The paper by Elperin and Gertsbakh (1987) demonstrates the implementation of the above method for confidence estimation.

3.3.5 Large-Sample Maximum Likelihood Confidence Intervals

The log-likelihood function (3.3.2) and the MLEs $\hat{\alpha}$ and $\hat{\beta}$ can be used for obtaining approximate confidence intervals for the parameters. First, we will describe the *modus operandi* of this approach. Later we will discuss the conditions under which this approach provides good results.

Suppose that we have at our disposal the MLEs $\hat{\alpha}, \hat{\beta}$.

1. Compute the following second-order derivatives:

$$V_{11}(\alpha, \beta) = -\frac{\partial^2 logLik}{\partial \alpha^2}, \tag{3.3.39}$$

$$V_{12}(\alpha, \beta) = -\frac{\partial^2 logLik}{\partial \alpha \partial \beta}, \tag{3.3.40}$$

$$V_{22}(\alpha, \beta) = -\frac{\partial^2 logLik}{\partial \beta^2}. \tag{3.3.41}$$

2. Substitute the MLEs into the expressions V_{ij}. Denote the corresponding expressions $\hat{V}_{ij} = V_{ij}(\hat{\alpha}, \hat{\beta})$. Form the following matrix called the ***Fisher observed information matrix***:

$$\mathbf{I}_\mathrm{F} = \begin{bmatrix} \hat{V}_{11} & \hat{V}_{12} \\ \hat{V}_{12} & \hat{V}_{22} \end{bmatrix}.$$

This matrix is positively definite.

3. Find the inverse of $\mathbf{I}_\mathrm{F}$:

$$\mathbf{I}_\mathrm{F}^{-1} = \begin{bmatrix} \hat{W}_{11} & \hat{W}_{12} \\ \hat{W}_{12} & \hat{W}_{22} \end{bmatrix}.$$

4. Let z_γ be the γ quantile of the standard normal distribution $N(0, 1)$. Approximate $1 - 2\gamma$ confidence intervals for α and β are calculated by the following formulas:

$$\text{for } \alpha: \ [\hat{\alpha} - z_\gamma \cdot \sqrt{\hat{W}_{11}},\ \hat{\alpha} + z_{1-\gamma} \cdot \sqrt{\hat{W}_{11}}], \tag{3.3.42}$$

$$\text{for } \beta: \ [\hat{\beta} - z_\gamma \cdot \sqrt{\hat{W}_{22}},\ \hat{\beta} + z_{1-\gamma} \cdot \sqrt{\hat{W}_{22}}]. \tag{3.3.43}$$

The principal statistical fact behind these formulas is the following: If the likelihood function has the form (3.3.1), where t_i are the observed values of i.i.d. random variables, and the density function $f(t;\alpha,\beta)$ satisfies some regularity conditions, the most critical of which is that the support does not depend on the parameters, then, as $n \to \infty$, the MLEs $\hat{\alpha}$ and $\hat{\beta}$ are asymptotically normal and unbiased, with variance–covariance matrix $\mathbf{I}_{\mathbf{F}}^{-1}$.

This result remains true if the observations are subject to type I and type II censoring. Under some additional conditions on $f(t;\alpha,\beta)$, it can be extended to the case of random independent censoring; see Lawless (1982, p. 526).

The crucial point is that the above result is true ***asymptotically***, i.e. for large samples. In reliability practice, a large sample of observed lifetimes is unlikely to be available. Applying the maximum likelihood method to small and medium-size samples, as well as to samples which are obtained under arbitrary censoring, may lead to inaccuracies in the actual confidence level. The issue of applicability of asymptotic results to finite samples needs further theoretical and experimental research. We suggest using the above described method with caution, and treating the results as approximate.

Example 3.3.3 continued: Large-sample confidence intervals for the parameters The necessary calculations are presented by the *Mathematica* printout in Fig. 3.4, which is a continuation of the printout in Fig. 3.3.

In[*6*] defines the elements of the matrix $\mathbf{I}_{\mathbf{F}}$ as the corresponding second-order partial derivatives. These derivatives are calculated by means of the operator "D[...]". Then the derivatives are numerically evaluated at the MLEs found in *Out*[4]; see *Out*[6], *Out*[7], *Out*[8]. *In*[9] defines the observed Fisher information matrix and its inverse as "Finv=Inverse[F]", and *Out*[10] is $\mathbf{I}_{\mathbf{F}}^{-1}$. *Out*[11] and *Out*[12] are the estimates of the corresponding standard deviations. We summarize this as follows: $\hat{\lambda} = 0.0137 \pm 0.0014$, $\hat{\beta} = 2.29 \pm 0.55$.

3.4 Exercises

1. Suppose $\tau_1 \sim W(\lambda,\beta)$, with $E[\tau_1] = 1$, and $Var[\tau_1] = 0.16$. Find λ and β. Suppose that $\tau_2 \sim logN(\mu,\sigma)$, with $E[\tau_2] = 1$ and $Var[\tau_2] = 0.16$. Find μ and σ. Compare the 0.1 quantiles of τ_1 and τ_2.

2. Mann and Fertig (1973) give the following data on a lifetime test of 13 identical components (in thousands of cycles): 0.22, 0.50, 0.88, 1.00, 1.32, 1.33, 1.54, 1.76, 2.50, 3.00. The test was terminated after the tenth failure.

a. Assume that the lifetime follows the Weibull distribution. Estimate parameters using the Weibull paper. Estimate $t_{0.1}$.

b. Solve the above problem for the lognormal distribution.
c. Find the MLE for Weibull parameters λ and β.

3. Ten items were put on test at $t = 0$ and inspected at $t = 2$. Four had failed, the rest were operating. Assume that the lifetime follows $W(\lambda = 0.4, \beta)$. Find numerically the MLE of β.

4. The lifetime τ is distributed as $\text{Exp}(a, \lambda)$; see (2.1.22). Suppose $t_1, \ldots, t_n$ are the observed lifetimes. Find the MLE of a and λ.
Hint: Derive from the log-likelihood that $\hat{a} = \min(t_1, \ldots, t_n)$.

5.The lifetime of a certain piece of equipment has c.d.f. $F(t; \alpha; \beta)$. n units operating at $t = 0$ were inspected at $t_1 = T$, $t_2 = 2T$ and $t_3 = 3T$. k_1 units failed in $[0, T]$, k_2 in $[T, 2T]$, and the remaining $n - k_1 - k_2$ failed in $[2T, 3T]$.
Write in a general form the expression for the likelihood function and for finding the MLE of α and β.

6. ***Replacement of a component on a finite interval*** $[0, T]$
a. Suppose a new component started operating at $t = 0$. When the component fails, it is immediately replaced by a new one. The replacements take place only on the interval $[0, T]$. The density function of component lifetime τ is $f(t; \alpha, \beta)$, and the c.d.f. is $F(t; \alpha, \beta)$.
Suppose the replacements took place at the instants $\{t_i\}$, $0 < t_1 < t_2 < \ldots < t_k < T$. Use this information to derive an expression for the likelihood function.

Hint: The following lifetimes were observed : $d_1 = t_1, d_2 = t_2 - t_1, \ldots, d_k = t_k - t_{k-1}$ and $\tau > T - t_k$. Thus the corresponding likelihood function is

$$Lik = \prod_{i=1}^{k} f(d_i; \alpha, \beta) \cdot [1 - F(T - t_k; \alpha, \beta)].$$

b. Assume that $\tau \sim F(t; \beta) = 1 - e^{-t^\beta}$. The following data were received in the replacement process described in part **a**: $T = 10, t_1 = 2, t_2 = 3.3, t_3 = 5.6, t_4 = 7.2, t_5 = 8.9$. Find the MLE of β.

7. ***The hazard plot***
a. If $R(t) = P(\tau > t)$ is the survival function, then the expression $\Lambda(t) = -\log R(t)$ is called the cumulative hazard.
Prove that if $\tau \in$ IFR, then the cumulative hazard is a convex function (assume that τ has a differentiable failure rate).
Hint. By (2.2.5), the cumulative hazard equals $\Lambda(t) = \int_0^t h(v)dv$. Differentiate

$\Lambda(t)$ twice, and use the property that for the IFR family, $h' > 0$.

b. The property established in **a** is used for a graphical investigation of $R(t)$. Plot $-\log \hat{R}(t_i)$ versus t_i, where $\hat{R}(t)$ is the Kaplan–Meier estimator of the survival function. This graph is called the ***hazard plot***. If the hazard plot reveals convexity, this might be an evidence that $\tau \in$ IFR.

Construct the hazard plot for the ABS data in Example 3.1.1.

```
In[6]:=  IF11 = -D[logLik, λ, λ] /. {λ → 0.0137, β → 2.292} // N
         IF12 = -D[logLik, λ, β] /. {λ → 0.0137, β → 2.292} // N
         IF22 = -D[logLik, β, β] /. {λ → 0.0137, β → 2.292} // N

Out[6]=  512978.

Out[7]=  180.019

Out[8]=  3.33224

In[9]:=  F = {{512978, 180.019}, {180.019, 3.33224}};
         FInv = Inverse[F]

Out[10]=  {{1.98707×10^-6, -0.000107348}, {-0.000107348, 0.305898}}

In[11]:=  sig[λ] = (1.98707*^-6)^0.5
          sig[β]  = (0.305898)^0.5

Out[11]=  0.00140963

Out[12]=  0.55308
```

Figure 3.4. *Mathematica* printout for the continuation of Example 3.3.3

Chapter 4

Preventive Maintenance Models Based on the Lifetime Distribution

When it is not necessary to change, it is necessary not to change
Viscount Lucius Cary Falkland

If it ain't broke, don't fix it

4.1 Basic Facts from Renewal Theory and Reward Processes

4.1.1 Renewal Process

When we choose a preventive maintenance scheme, we are usually interested in selecting the maintenance parameters, e.g. the maintenance period, in the "best" possible way. To do this, we need to compare the expressions for mean (expected) *costs* or *rewards* for various maintenance periods. Our first task is to learn how to write expressions for these costs or rewards. We will need some basics from the renewal theory.

Definition 4.1.1: *Counting process*
$\{N(t),\ t > 0\}$ is said to be a counting process if $N(t)$ represents the total

number of "events" on $(0, t]$.

Obviously, $N(t)$ is integer-valued, $s < t$ implies that $N(s) \le N(t)$, $N(0) = 0$, and $N(t) - N(s)$ is the number of events on $(s, t]$.

Denote by X_n the random time between the $(n-1)$th and nth events; see Fig. 4.1. (The zeroth event takes place at $t = 0$ and is ignored.)

Definition 4.1.2: *Renewal process*
Let $\{X_i\}$ be i.i.d. nonnegative random variables, $X_i \sim F(t)$. Then the counting process $N(t)$

$$N(t) = \{\max \; n : S_n = X_1 + X_2 + \ldots + X_n \le t\} \tag{4.1.1}$$

is called a *renewal process.*

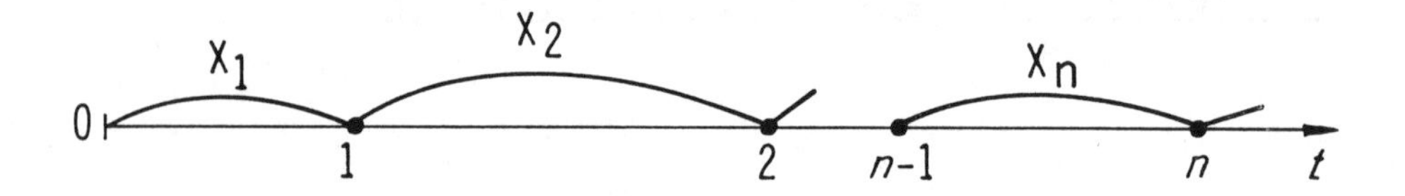

Figure 4.1. The renewal process

To put it simply, the renewal process counts the number of X_i intervals on $(0, t]$. If $E[X_i] = \mu$, then by the law of large numbers, $S_n/n \to \mu$ as $n \to \infty$ with probability 1.

Let us establish the distribution of $N(t)$. Write $F_n(t) = P(S_n \le t)$. Denote $F_1(t) = F(t)$, the c.d.f. of X_i.

Theorem 4.1.1
$P(N(t) = n) = F_n(t) - F_{n+1}(t)$.

Proof
$N(t) \ge n \Leftrightarrow S_n \le t$. Then $P(N(t) = n) = P(S_n \le t) - P(S_{n+1} \le t)$.

Definition 4.1.3: *Renewal function*
The mean number of events $m(t)$ on $(0, t]$ is called the *renewal function*: $E[N(t)] = m(t)$.

Theorem 4.1.2

$$m(t) = \sum_{n=1}^{\infty} F_n(t) \, . \tag{4.1.2}$$

Proof
$m(t) = \sum_{n=1}^{\infty} nP(N(t) = n)$, which after some algebra, is equal to $F_1(t) + F_2(t) + \ldots + F_n(t) + \ldots$.

Example 4.1.1: Poisson process
In a Poisson process, $X_i \sim \text{Exp}(\lambda)$; see Sect. 2.1. From the description and the properties of this process we know that the number of "calls" (events) on $(0, t]$ has a Poisson distribution with parameter λt. Therefore, the mean number of events on $(0, t]$ equals λt and thus $m(t) = \lambda t$.

Theorem 4.1.3
As $t \to \infty$, with probability 1,

$$\frac{N(t)}{t} \to \frac{1}{\mu}. \tag{4.1.3}$$

Proof (Ross 1993)
Obviously

$$S_{N(t)} \le t < S_{N(t)+1}, \tag{4.1.4}$$

which we can also express as

$$\frac{S_{N(t)}}{N(t)} \le \frac{t}{N(t)} < \frac{S_{N(t)+1}}{N(t)}. \tag{4.1.5}$$

As $t \to \infty$, so does $N(t)$. $S_{N(t)}/N(t)$ is an average of $N(t)$ i.i.d. random variables $X_1, \ldots, X_{N(t)}$, and by the strong law of large numbers, $S_{N(t)}/N(t) \to \mu$. Write

$$S_{N(t)+1}/N(t) = \Big(S_{N(t)+1}/(N(t)+1)\Big)\Big((N(t)+1)/N(t)\Big).$$

The first factor tends to μ and the second to 1, as t goes to infinity. Therefore, $t/N(t) \to \mu$, or $N(t)/t \to 1/\mu$. The quantity $1/\mu$ is called the *renewal rate*.

Theorem 4.1.4: *Elementary renewal theorem*
As $t \to \infty$,

$$\frac{m(t)}{t} \to \frac{1}{\mu}. \tag{4.1.6}$$

We omit the proof.

At first glance, this theorem seems to be a corollary of Theorem 4.1.3 since the convergence of $N(t)/t$ to $1/\mu$ must imply that the $E[N(t)]/t$ also converges to $1/\mu$. There are, however, some subtle details which complicate the proof; see Ross (1970, p. 40) or Ross (1993, p. 275).

Theorem 4.1.5
Suppose the inter-renewal intervals have mean μ and variance σ^2. Then, as $t \to \infty$,

$$[m(t) - t/\mu] \to \frac{\sigma^2}{2\mu^2} - \frac{1}{2}\,. \tag{4.1.7}$$

We omit the proof, which can be found in Gnedenko et al (1969). The theorem states that $m(t)$ has an asymptote, as shown in Fig. 1.2.

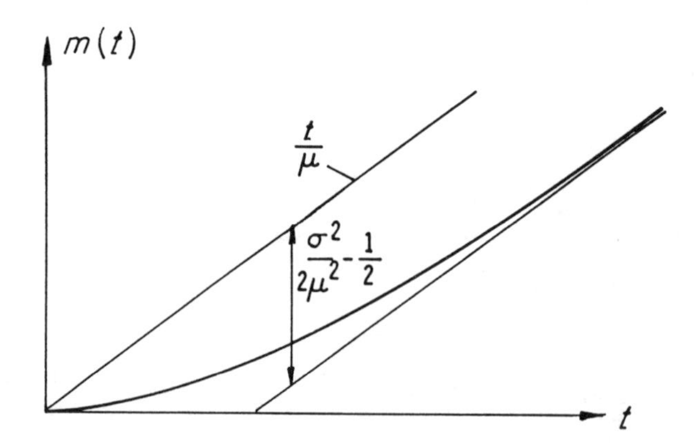

Figure 4.2. The behavior of the renewal function $m(t)$

Theorem 4.1.6: *Asymptotic distribution of* $N(t)$
As $t \to \infty$,

$$\frac{N(t) - t/\mu}{\sqrt{t\,\sigma^2/\mu^3}} \xrightarrow{d} N(0,1). \tag{4.1.8}$$

Loosely speaking, $N(t)$ for large t is approximately normal with mean t/μ and variance $t\sigma^2/\mu^3$. We omit the proof, which can be found in Feller (1968, Chap. 13).

In applications, we often need an explicit expression for the renewal function $m(t)$. Unfortunately, closed-form expressions for $m(t)$ can be found only for the gamma and normal cases.

Example 4.1.2: m(t) for gamma and normal cases
(i) *The gamma distribution.* Let $X_i \sim Gamma(k, \lambda)$. Note that $F_n(t)$ is an n-fold convolution of this distribution. Remembering the model of the gamma distribution, $F_n \sim Gamma(nk, \lambda)$. Therefore, writing $F_1(t) = Gamma(t;\ k, \lambda)$,

$$m(t) = \sum_{n=1}^{\infty} Gamma(t;\ kn, \lambda)\,. \tag{4.1.9}$$

(ii) *The normal distribution.* Assume that the normal distribution is used to describe a nonnegative random variable. This implies that the negative tail might be ignored. Then

$$m(t) = \sum_{n=1}^{\infty} \Phi\Big(\frac{t - n\mu}{\sigma\sqrt{n}}\Big) . \tag{4.1.10}$$

However, for most cases we are forced to use approximations to the renewal function. It is easy to show that the nth convolution is always less than or equal to $[F_1(t)]^n$: $F_n(t) \le [F(t)]^n$, where we put $F_1(t) = F(t)$. Then it follows from Theorem 4.1.2 that

$$F_1(t) + \ldots + F_n(t) \le m(t) \le F_1(t) + \ldots + F_n(t) + \sum_{i=n+1}^{\infty} [F(t)]^i . \tag{4.1.11}$$

For practical purposes, for small t values, it is enough in (4.1.11) to take $n = 2$. Then, assuming that $F(t) < 1$,

$$F_1(t) + F_2(t) < m(t) < F_1(t) + F_2(t) + [F(t)]^3/(1 - F(t)) . \tag{4.1.12}$$

This approximation is reasonably good for $t < 0.5E[X_i]$.

To facilitate the use of the renewal function in computations related to preventive maintenance, we present in Appendix E a table of the renewal function for the Weibull family. The results were obtained by using the approximation $m(t) \approx \sum_{n=1}^{5} F_n(t)$. We used the following recursive formula:

$$F_n(t) = \int_0^t f(x) F_{n-1}(t - x) dx,$$

where $F_1(t) = F(t) = 1 - \exp(-t^\beta)$, $f(t) = dF(t)/dt$. The integral has been replaced by a finite sum. The maximum error of the approximation in the table does not exceed 0.001.

The following theorem proved by D. Blackwell describes the local behavior of the renewal function $m(t)$ as t goes to infinity.

Theorem 4.1.7 (Blackwell)
Let the c.d.f. of X_i be a continuous function. Then, for all $a > 0$,

$$\lim_{t\to\infty} [m(t + a) - m(t)] = \frac{a}{\mu} . \tag{4.1.13}$$

We omit the proof, which can be found in Ross (1970, p. 41). From (4.1.13) it follows that $\lim_{t\to\infty}[m(t + a) - m(t)]/a = 1/\mu$ and thus

$$\lim_{a\to 0} \lim_{t\to\infty} \frac{m(a + t) - m(t)}{a} = \frac{1}{\mu} . \tag{4.1.14}$$

Suppose it is possible to interchange the limits. Then

$$\lim_{t\to\infty} \frac{dm(t)}{dt} = m'(t) = \frac{1}{\mu}, \tag{4.1.15}$$

i.e. the derivative of the renewal function for "large" t equals the renewal rate. $m'(t)$ is called the ***renewal density*** or the ***renewal rate.***

Let us give a probabilistic interpretation of the renewal density. Obviously

$$\begin{aligned} m'(t)\Delta t &\approx m(t+\Delta t) - m(t) \\ &= E[\#\text{ of renewals in } (t, t+\Delta t)] \\ &= P(\text{exactly one renewal in } (t, t+\Delta t)) + \\ &\quad \sum_{j>1} jP(j \text{ renewals in } (t, t+\Delta t)) . \end{aligned}$$

It can be shown (see Gnedenko et al 1969, Sect. 2.3) that the sum in the last formula is $o(\Delta t)$ as $\Delta t \to 0$. Thus,

$$m'(t)\cdot\Delta t \approx P(\text{exactly one renewal in } (t, t+\Delta t)). \tag{4.1.16}$$

4.1.2 Renewal Reward Process

Let us associate with each renewal period a nonnegative random variable R_i called the *reward* or *cost.* It will be assumed that $\{R_i\}$ are i.i.d. random variables, with mean value $E[R_i] = E[R]$.

Let us assume that the reward R_i for the ith renewal period is accounted at the end of this period. Then the total reward accumulated on the interval $[0, t]$ will be $R(t) = \sum_{i=1}^{N(t)} R_i$, since there are $N(t)$ renewal periods on $(0, t]$.

The following theorem is the key to computing the reward per unit time.

Theorem 4.1.8

(i) With probability $\to 1$,

$$\frac{R(t)}{t} \to \frac{E[R]}{\mu} \tag{4.1.17}$$

as $t \to \infty$.

(ii) As $t \to \infty$,

$$\frac{E[R(t)]}{t} \to \frac{E[R]}{\mu} . \tag{4.1.18}$$

Proof

(i) Write

$$\frac{R(t)}{t} = \frac{\sum_{i=1}^{N(t)} R_i}{t} = \frac{\sum_{i=1}^{N(t)} R_i}{N(t)} \frac{N(t)}{t} . \tag{4.1.19}$$

By the strong law of large numbers, the first fraction on the right-hand side tends to $E[R]$ as $t \to \infty$. By Theorem 4.1.3, the second ratio tends to $1/\mu$, which proves (i).
(ii) It might seem strange that here the proof is more difficult than for (i). We omit it and refer the reader to Ross (1993).

Example 4.1.3: Comparing two replacement strategies
A piece of equipment has two independent parts, a and b. Their lifetimes are r.v.s X_a and X_b, respectively: $X_a \sim F(t)$ and $X_b \sim G(t)$. Let $E[X_a] = \mu_a$, $E[X_b] = \mu_b$.

If a part fails, an emergency repair (ER) takes place: the part is immediately replaced by an equivalent new one. The cost of an ER is $c_{ER} = \$1000$, a "large" cost. There is also an option to combine the ER of one part with a preventive repair (PR) of the another one. PR means replacing the part in the absence of failure by an equivalent new one. The cost of *combined* repair is $c_{com} = c_{ER} + \Delta$, and Δ is relatively small. Assume $c_{com} = \$1030$.

Two service strategies are suggested: replace each part at its failure only (strategy I); together with the ER of one part, carry out the PR of the other one (strategy II). Strategy II is often called ***opportunistic***. It is decided to compare both strategies by comparing the values of the expected cost per unit time over an infinite time period.

Strategy I. Obviously the cost per unit time η_I is

$$\eta_I = \lim_{t\to\infty} E\Big[\text{the cost over } [0,t]\Big]/t = c_{ER}(1/\mu_a + 1/\mu_b) \,. \tag{4.1.20}$$

Strategy II. Let $Z = \min(X_a, X_b)$, and let $E[Z] = \mu_0$. Then, on the average, each μ_0 units of time, a cost c_{com} is paid. Thus the cost per unit time is $\eta_{II} = c_{com}/\mu_0$.

Let us compare these strategies for two cases : X_a and X_b both exponential, with parameters λ_a and λ_b, respectively; and X_a and X_b both Weibull, with parameters $\lambda = 1, \beta = 2$ for X_a, and $\lambda = 2, \beta = 2$ for X_b.

For the first case, $Z = \min(X_a, X_b) \sim \text{Exp}(\lambda_a + \lambda_b)$ and thus $\eta_I = c_{ER} \times (\lambda_a + \lambda_b)$. As to η_{II}, it equals $c_{com}(\lambda_a + \lambda_b)$.

So, the conclusion is that for exponential lifetimes, strategy I is always preferable. It is, in fact, clear without any formal derivation because for an exponential (nonaging) part a preventive replacement (even for a small extra cost) is just a waste of money.

For the Weibull case the situation is different. First, let us compute the means of X_a and X_b. The easiest way is to use the formula $E[X] = \int_0^\infty (1 - F(t))dt$ which gives $\mu_a = 0.886$, and $\mu_b = 0.443$. One can use numerical integration in *Mathematica*.

To compute $E[Z]$, use (1.3.4) and (1.3.12).

$$P(Z > t) = \exp[-(\lambda_a t)^\beta] \exp[-(\lambda_b t)^\beta] = \exp[-(\sqrt{5}t)^2].$$

Now $E[Z] = \int_0^\infty P(Z > t)dt = 0.396$, and $\eta_I = \$3390$ and $\eta_{II} = \$2600$. Therefore, the opportunistic policy is better than the replace-only- upon-failure policy.

4.1.3 Alternating Renewal Process. Component Stationary Importance Measure for a Renewable System

Consider a component which has alternating "up" and "down" states. Assume that the up periods are described by a sequence of i.i.d. positive random variables $Y_j, j = 1, 2, \ldots$, and the down periods by another sequence of positive i.i.d. random variables $X_j, j = 1, 2, \ldots$. These two sequences are assumed to be independent. It is convenient to assume that time $t = 0$ coincides with the end of an up period, and thus the duration of alternating states is described by the sequence $X_1, Y_1, X_2, Y_2, \ldots$.

Let $Z_i = X_i + Y_i$, $i = 1, 2, \ldots$. These r.v.s describe statistically identical operation cycles. At the instants $Z_1, Z_1 + Z_2, \ldots, Z_1 + \ldots + Z_k, \ldots$, the component fails. Obviously, the counting process $N_a(t) = \{\max n : S_n = Z_1 + Z_2 + \ldots + Z_n \le t\}$ is a renewal process.

Let $E[X_j] = \nu$, $E[Y_j] = \mu$. Obviously, $\nu + \mu$ is the mean interval between renewals in the process $N_a(t)$. Let $A_v(t)$ be the ratio:

$$A_v(t) = \frac{\text{total time up on } [0, t]}{t}. \tag{4.1.21}$$

Obviously the numerator lies between $Y_1 + \ldots + Y_{N_a(t)}$ and $Y_1 + \ldots + Y_{N_a(t)+1}$.

We are interested in the behavior of $A_v(t)$ as $t \to \infty$. Let us divide the numerator and the denominator by $N_a(t)$. As $t \to \infty$, $N_a(t)$ also goes to infinity. Since $(Y_1 + \ldots + Y_{N_a(t)})/N_a(t) \to \mu$, and $t/N_a(t) \to (\nu + \mu)$, (see Theorem 4.1.3), we arrive at the following result:

$$\lim_{t\to\infty} A_v(t) = \frac{\mu}{\mu + \nu} = A_v. \tag{4.1.22}$$

This limit is called ***stationary availability***. This notion was introduced in Section 1.2. The probabilistic interpretation of A_v is as the probability that at some remote instant t the component is in the up state.

The quantity $1/(\mu + \nu)$ is the limit of the renewal density as $t \to \infty$. We call this limit the ***stationary renewal rate.***

Recalling the probabilistic interpretation of the renewal density (see above), $\Delta t/(\mu + \nu)$ is the probability that the component fails in the interval $(t, t + \Delta t)$, as t tends to infinity.

Now suppose that we have a monotone system consisting of n ***independent*** renewable components. Each component has alternating up and down intervals. To describe the renewal process for component j, we use the previously introduced notation and add index j. So the stationary availability of component j is denoted by $A_v(j)$ and equals $\mu(j)/(\mu(j) + \nu(j))$.

Suppose also that we know the system structure function $\phi(x_1, \ldots, x_n)$. Recall from Chap. 1 that $E[\phi(X_1, \ldots, X_n)] = \psi(A_v(1), \ldots, A_v(n)) = A_v$ is the system stationary availability.

From time to time the system fails. We are interested in obtaining the relative proportion of system failures coinciding with the ***failure of component*** j.

Denote by $Q(t)$ the total number of system failures in the interval $(0, t]$ and by $Q_j(t)$ the total number of system failures caused by the failure of component j. We call the limiting value of the fraction $Q_j(t)/Q(t)$ (for $t \to \infty$) ***Barlow's stationary index of importance of component*** j and denote it by B_j:

$$B_j = \lim_{t \to \infty} \frac{Q_j(t)}{Q(t)}. \tag{4.1.23}$$

Let us consider an example which clarify the meaning of B_j. Exercise 1 in Sect. 1.4 considers a series-parallel system of four components: components 2 and 3 are in parallel, both are in series with component 1, and components 1, 2, 3 are in parallel with component 4. Suppose we are interested in system failures caused by the failures of component $j = 2$. Its stationary renewal rate equals $1/(\mu(2) + \nu(2))$. Not each failure of this component causes system failure. For example, let component 2 fail when 1 and 3 are up and 4 is down. The system will not fail. Similarly, if 1 and 4 are down and component 2 fails, it does not cause the system to fail because it is already in the down (failure) state. On the other hand, let component 2 fail when 3 and 4 are down and 1 is up. Then clearly this failure causes system failure.

More formally, we might say that the failure of component 2 coincides with system failure if $\phi(x_1, 1_2, x_3, x_4) - \phi(x_1, 0_2, x_3, x_4) = 1$. Indeed, this difference equals 1 if and only if the first term is 1 and the second is 0, which means that
(i) the system with component 2 in the up state is up *and*
(ii) the system with component 2 in the down state is down.

Since $\phi(\mathbf{X})$ is a binary variable,

$$\begin{aligned} &P(\phi(X_1, \ldots, 1_j, \ldots, X_n) - \phi(X_1, \ldots, 0_j, \ldots, X_n) = 1) \\ &= E[\phi(X_1, \ldots, 1_j, \ldots, X_n) - \phi(X_1, \ldots, 0_j, \ldots, X_n)] \\ &= E[\phi(X_1, \ldots, 1_j, \ldots, X_n)] - E[\phi(X_1, \ldots, 0_j, \ldots, X_n)] \\ &= \psi(A_v(1), \ldots, 1_j, \ldots, A_v(n)) - \psi(A_v(1), \ldots, 0_j, \ldots, A_v(n) . \end{aligned} \tag{4.1.24}$$

The first term on the last line is the stationary availability of the system in which the component j is "absolutely reliable", i.e. is always up. The second term is the stationary availability of the system with component j always being down. Note that the expression in (4.1.24) does not involve the reliability characteristics of component j .

Note also that (4.1.24) is Birnbaum's importance measure of component j defined as $\partial\psi/\partial p_j$ (see Exercise 2 in Sect. 1.4). Indeed, by pivoting around component j, one obtains that

$$\psi(A_v(1),\ldots,A_v(n)) = A_v(j)\psi(A_v(1),\ldots,1_j,\ldots,A_v(n)) +(1-A_v(j))\psi(A_v(1),\ldots,0_j,\ldots,A_v(n)). \quad (4.1.25)$$

Now

$$\frac{\partial\psi(\cdot)}{\partial A_v(j)} = \psi(A_v(1),\ldots,1_j,\ldots,A_v(n)) - \psi(A_v(1),\ldots,0_j,\ldots,A_v(n)) . \quad (4.1.26)$$

Now it is clear that Birnbaum's importance measure of component j applied to component stationary probabilities $p_j = A_v(j)$ is the stationary probability that the system is in the "j-critical state", i.e. in such a state that component j's failure coincides with system failure. On a "large" time span $[0,t]$, the mean duration of the time during which the system is j-critical equals $t\cdot\partial\psi(A_v(1),\ldots,A_v(n))/\partial A_v(j)$.

Now let us consider a small time interval $(t, t+\Delta t)$ as $t\to\infty$. The event C_j = "component j fails in this interval *and* the system is in j-critical state" has probability

$$P(C_j) \approx \frac{\partial\psi(A_v(1),\ldots,A_v(n))}{\partial A_v(j)} \frac{\Delta t}{\mu(j)+\nu(j)}. \quad (4.1.27)$$

The above reasoning is in fact a heuristic proof of the following theorem by Barlow (1998, Sect. 8.3):

Theorem 4.1.9

$$\begin{aligned} B_j &= \lim_{t\to\infty} \frac{Q_j(t)}{\sum_{i=1}^n Q_i(t)} \\ &= \frac{(\mu(j)+\nu(j))^{-1}(\partial\psi(A_v(1),\ldots,A_v(n)))/\partial A_v(j)}{\sum_{i=1}^n(\mu(i)+\nu(i))^{-1}(\partial\psi(A_v(1),\ldots,A_v(n))/\partial A_v(i))}. \end{aligned} \quad (4.1.28)$$

Example 4.1.4
Compute Barlow's stationary importance measures B_j for the system described in Exercise 1, Sect. 1.4, for the following data: $\mu(1)=1$, $\nu(1)=0.2$; $\mu(2)=1$, $\nu(2)=0.3$; $\mu(3)=0.8$, $\nu(3)=0.2$; $\mu(4)=1.5$, $\nu(4)=0.5$.

The stationary availabilities are: $A_v(1)=0.833$; $A_v(2)=0.769$; $A_v(3)=0.8$, $A_v(4)=0.75$. The structure function of the system is

$$\begin{aligned} \phi(x_1,\ldots,x_4) &= 1-(1-x_1x_2)(1-x_1x_3)(1-x_4) \\ &= x_1x_2+x_1x_3-x_1x_2x_3+x_4-x_1x_2x_4-x_1x_3x_4 \\ &+ x_1x_2x_3x_4. \end{aligned}$$

Replace x_i by $A_v(i)$ and take the partial derivatives:

$$\begin{aligned}\partial\psi(\cdot)/\partial A_v(1) &= A_v(2) + A_v(3) - A_v(2)A_v(3) - A_v(2)A_v(4) \\ &- A_v(3)A_v(4) + A_v(2)A_v(3)A_v(4) = 0.238;\end{aligned}$$

$$\begin{aligned}\partial\psi(\cdot)/\partial A_v(2) &= A_v(1) - A_v(1)A_v(3) - A_v(1)A_v(4) + A_v(1)A_v(3)A_v(4) \\ &= 0.042;\end{aligned}$$

$$\begin{aligned}\partial\psi(\cdot)/\partial A_v(3) &= A_v(1) - A_v(1)A_v(2) - A_v(1)A_v(4) + A_v(1)A_v(2)A_v(4) \\ &= 0.048;\end{aligned}$$

$$\begin{aligned}\partial\psi(\cdot)/\partial A_v(4) &= 1 - A_v(1)A_v(2) - A_v(1)A_v(3) + A_v(1)A_v(2)A_v(3) \\ &= 0.205.\end{aligned}$$

Substituting these values and the values of $(\mu(i) + \nu(i))^{-1}$ into (4.1.27), we obtain that $B_1 = 0.52$; $B_2 = 0.08$; $B_3 = 0.13$; $B_4 = 0.27$. Thus, 52% of system failures coincide with the failures of component 1.

4.2 Principal Models of Preventive Maintenance

The models which we describe now are central to the applications of preventive maintenance theory. We will describe the preventive maintenance policy and derive an expression for a cost (reward) criterion to characterize each model.

4.2.1 Periodic (Block) Replacement – Cost-type Criterion

A new unit starts operating at $t = 0$. At each of the time instants T, $2T$, $3T, \ldots$ the unit is replaced by a new one, from the same population. This replacement, termed *preventive maintenance* (PM), costs c, where $c < 1$. At each failure which appears between the PMs, the unit is also replaced by a new one. This replacement upon failure is called an *emergency repair* (ER) and costs $c_{ER} = 1$. All replacements take negligible time. The information available is the c.d.f. of unit lifetime $F(t)$. Figure 4.3 explains the block replacement model.

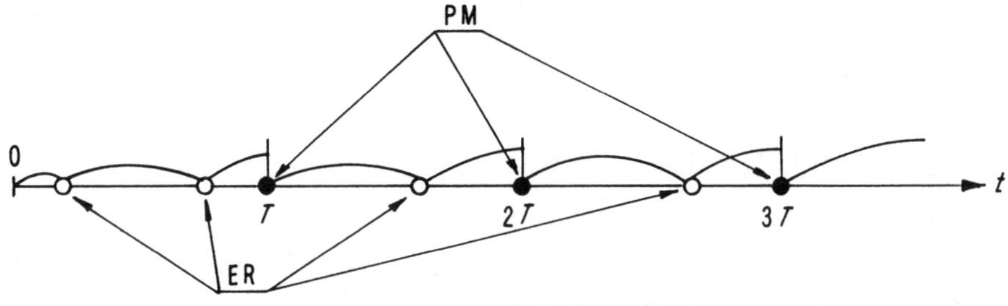

Figure 4.3. Scheme of block replacement

Here the ERs between two adjacent PMs form a renewal process on $[0, T]$. The mean cost over one period of length T equals $E[R_i] = c + m(T) \times 1$, where $m(T)$ is the mean number of ERs on $[0, T]$. Then the mean cost per unit time is

$$\eta_A(T) = \frac{c + m(T)}{T}. \tag{4.2.1}$$

Block replacement is the simplest and most widely used maintenance scheme. A formal complication with the expression for $\eta_A(T)$ is that it depends on the renewal function, which is difficult to compute. Often the following upper and lower bounds are satisfactory (see (4.1.12)):

$$\frac{c + F(T) + F^2(T)}{T} < \eta_A(T)$$
$$< \frac{c + F(T) + F^2(T) + [F(T)]^3/(1 - F(T))}{T}. \tag{4.2.2}$$

Investigation of the expression for $\eta_A(T)$ is postponed until later. Obviously, small values of T lead to large costs for frequent PMs. Large values of T save on PMs but lead to large costs on ERs.

It is worth noting that as $T \to 0$, $\eta_A(T) \to \infty$; if $T \to \infty$, then by Theorem 4.1.4 $m(T)$ behaves as T/μ and $\eta_A(T) \to 1/\mu$.

Remark: the relevant time scale

It is important to determine what is meant by "time" in the context of preventive maintenance in general, and in the context of block replacement in particular. Time t may have several meanings, depending on the specific circumstances: calendar time, operation time, the number of operation cycles, etc. If the time is defined as *calendar* time, then the c.d.f. of the system lifetime is related to calendar time and the maintenance periods are expressed as calendar units, say hours, days or months. If the relevant time is the *operational* ("up") time, then the lifetime distribution, preventive maintenance periods, etc. are measured in corresponding units, e.g. operation hours. In many cases, the choice of the relevant time units is obvious. For example, calendar time is a natural choice for continuously operating equipment. Very often, there might be several competing "parallel" time scales, such as the calendar time and the mileage scales for a car. It is not immediately clear which of these two scales is the "true" one for each particular car system. We will return to the choice of the best time scale in Chap. 6.

4.2.2 Block Replacement: Availability Criterion

A new unit starts operating at $t = 0$. The time axis is calendar time, which includes operation and idle time. At each failure, an ER is carried out which lasts time t_{ER}. After the total accumulated *operational* time reaches T, a PM is carried out, which takes t_{PM}. Both ER and PM completely renew the unit.

After a PM, the process repeats itself, as is seen from Fig. 4.4. Typically, $t_{PM} \ll t_{ER}$.

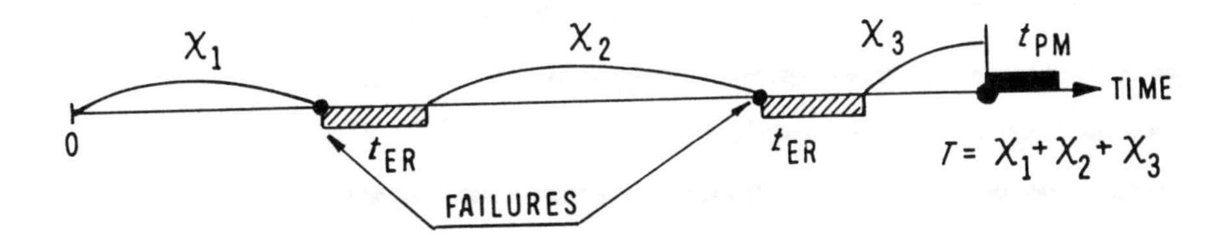

Figure 4.4. PM is carried out after each T units of operational time

We are interested in maximizing the ***stationary availability***. One renewal period has the mean duration $T + m(T) \cdot t_{ER} + t_{PM}$. The total operational ("up") time on this period equals T. This is the "reward" in our situation. The average reward per unit of calendar time gives the stationary availability:

$$\eta_B(T) = \frac{T}{T + m(T)t_{ER} + t_{PM}} .$$

Divide the numerator and denominator by T. Then this expression takes the form

$$\eta_B(T) = \frac{1}{1 + t_{ER}\eta_A^*(T)}, \tag{4.2.3}$$

where $\eta_A^*(T)$ is the cost for block replacement with $c = t_{PM}/t_{ER}$. So, maximizing the availability is equivalent to minimizing the corresponding costs.

4.2.3 Periodic Group Repair – Operation Time Based Criterion

A group of n cold drinks machines located in the same building are serviced according to the following rule. Each T units of calendar time they are visited by a technician who checks all the machines, loads them and eliminates all malfunctions. This work completely "renews" all the machines. The cost of the technician's visit is, on average, $c_p + nc_0$. The duration of the service is assumed to be negligible (it can be done, say, at night when the machines are not in use). Each machine provides a revenue of $\$r_1$ per unit of its operation time and a loss, i.e. a negative "revenue," $\$r_2$ for each unit of its idle time, $r_2 < 0$. Figure

4.5 illustrates the operation and servicing of the machines.

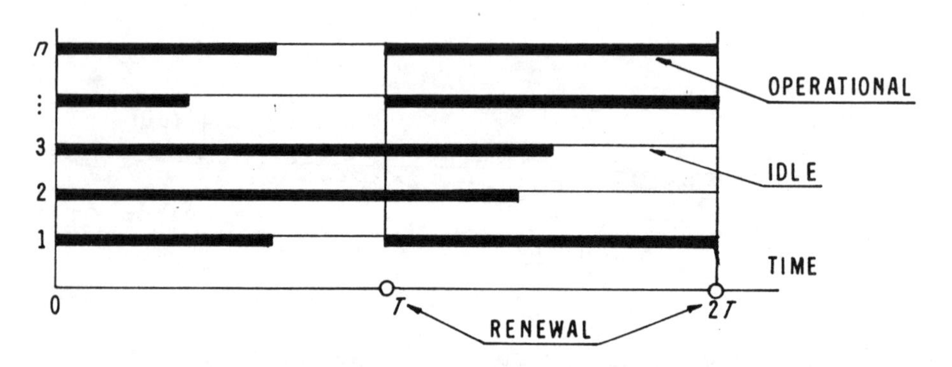

Figure 4.5. Periodic servicing of n machines

We assume that we know the distribution of the operational time τ for each machine: $\tau \sim F(t)$. Let $f(t)$ be the corresponding density function.

Let us find the expression for the mean reward per unit of calendar time. Note that we have a renewal process with fixed renewal period T. A machine fails in the interval $[x, x+dx]$, with probability $f(x)dx$, for $x \in [0, T]$ and does not fail in $[0, T]$ with probability $1 - F(T)$. Thus the mean reward from one machine on the interval $[0, T]$ is

$$r(T) = \int_0^T (r_1 x + r_2(T - x))f(x)dx + r_1 T(1 - F(T)) \,. \tag{4.2.4}$$

Now the reward per unit time from n machines is obviously

$$\eta_C(T) = \frac{nr(T) - c_p - nc_0}{T}. \tag{4.2.5}$$

The expression for $\eta_C(T)$ can be further simplified. It is advisable to carry out its investigation numerically. There is typically an optimal period T^* which maximizes the reward, if r_2 is negative: if T is very small, frequent visits by the technician will be very costly; if T is very large, then most of the time the machines will be idle. It is possible to establish that $\lim \eta_C(T) = -\infty$ as $T \to 0$ and that $\lim \eta_C(T) = nr_2$ as $T \to \infty$.

Example 4.2.1
Assume $r_1 = 1$, $r_2 = -0.1$, $F(t) = 1 - \exp[-t]$. The repair costs are $c_p = 0.1, c_0 = 0.01$ and the number of machines in the group is $n = 10$.

Figure 4.6 shows the graph of $\eta_C(T)$. It has a maximum at $T^* \approx 0.2$, which is about 1/5 of the machine mean up time. The maximal $\eta \approx 8$. It is instructive to note that very frequent technician's visits, say with $T = 0.15$

would considerably lower the reward per unit time, while taking $T = 0.5$ of the mean up time leads to much smaller decrease in the η_C value.

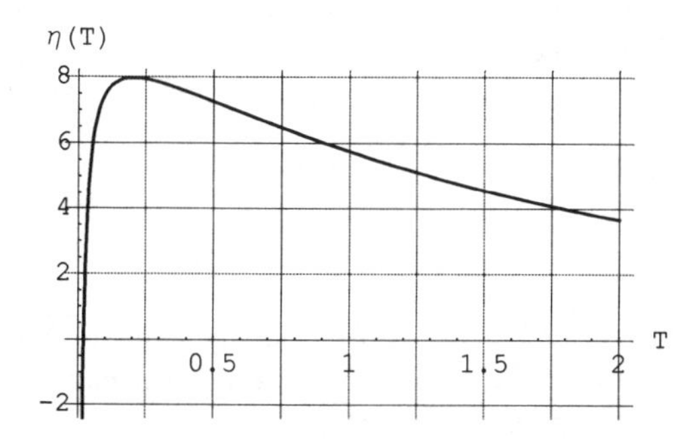

Figure 4.6. $\eta_C(T)$ in Example 4.2.1

Another interesting feature of the maintenance model considered in this subsection is that there is an optimal preventive maintenance policy even if the lifetime follows the exponential distribution, as the graph shows! On the one hand, there is no need for PM if all machines are up, if the lifetimes are exponential. But on the other hand, in most cases, the PMs put an end to the idle period and thus increase the rewards.

4.2.4 Periodic Preventive Maintenance with Minimal Repair

Minimal repair with complete periodic renewal

The weakness of all preventive maintenance models previously considered was the assumption of a *complete* renewal at failure and at the PM. The present model makes a more realistic assumption: the ER eliminates the failure but does not change the failure rate. Recall that the failure rate $h(t)$ has the following meaning: $h(t)\Delta t \approx P(\text{failure in } (t, t+\Delta t)|\tau > t)$.

The *minimal* repair made at time t_0 eliminates the failure but leaves $h(t_0)$ *unchanged.*

An important fact is the following.

Theorem 4.2.1

Under minimal repair, the mean number of failures on $[0, T]$ is equal to

$$H(T) = \int_0^T h(t)dt\,. \tag{4.2.6}$$

We omit the proof, which is based on the fact that for minimal repair the failure instants follow a Poisson process with *time-dependent* event rate $\lambda(t) = h(t)$; see Appendix A.

In particular, the probability of absence of failures on $[0, T]$ equals $e^{-H(T)}$. This formula is similar to (2.1.3).

If ER costs $c_{ER} = 1$ and PM costs $c_{PM} = c$, then the average cost per unit time is

$$\eta_D(T) = \frac{H(T) + c}{T} . \tag{4.2.7}$$

Example 4.2.2
Let $h(t) = \lambda$. Then $\eta_D(T) = \lambda + c/T$, and obviously the best strategy is $T = \infty$, i.e. to avoid any PM. This is quite expected because a constant failure rate characterizes the exponential distribution.

Suppose now that $h(t) = at$ (this corresponds to the Weibull distribution with shape parameter $\beta = 2$). It is easy to derive that $\eta_D(T) = aT/2 + c/T$. The optimal T^* always exists and is equal to $T^* = \sqrt{2c/a}$.

Minimal repair with partial renewal

Complete renewal, as considered above, demands in practical terms either replacement of the whole component (or the system) by a new one, or a series of repair actions which would bring each part of the system to a "brand new" state. For example, all worn mechanical parts would be replaced by new ones. This is not always technically feasible. Very often, periodic repair activity improves the system but does not bring it to the brand new state (we call this "partial renewal").

It will now be assumed that this partial repair (e.g. lubrication, replacement of badly worn parts, adjustment and tuning of others) made at instant $t = T$ does not bring the system failure rate $h(t)$ to its initial level $h(0)$. Formally, the behavior of the failure rate on the interval $[T, 2T]$ will *not* be a copy of the failure rate behavior on the interval $[0, T]$.

In the literature, several models of partial renewal have been proposed. Zhang and Jardine (1998) suggest that partial renewal makes the system failure rate between "bad as old" and "good as new", which means the following. System failure rate at $t = 0$ is $h(0)$; before the partial repair which is carried out at time t_0, system failure rate is $h(t_0)$. The partial repair means reducing the failure rate to the value $h(t^*)$ which lies between $h(0)$ and $h(t_0)$. The paper by Usher et al (1998) considers a partial renewal which reduces the actual system age by a certain fraction.

We suggest the following *partial renewal* model. In each interval $I_k = [T(k-1), Tk]$, the system failure rate will be equal to the system failure rate on the previous interval I_{k-1} multiplied by a "degradation" factor e^{α}, where $\alpha > 0$ and α is an unknown parameter. For example, if the failure rate in I_1 is $h(t)$ and ranges between $h(0)$ and $h(T)$, then in I_2, after the partial renewal at $t = T$, the failure rate will be $e^{\alpha}h(t)$ ranging between $h(0)e^{\alpha}$ and $h(T)e^{\alpha}$. In the third interval I_3, the failure rate will vary between $e^{2\alpha}h(0)$ and $e^{2\alpha}h(T)$, etc. This implies that the mean number of failures in interval I_1 is $H_1 = H(T)$, in I_2 $H_2 = e^{\alpha}H(T)$, and so on. In I_k, the mean number of failures will be equal to $H_k = e^{(k-1)\alpha H(T)}$.

In our calculations, only the mean number of failures H_k in the intervals $I_k, k \geq 1$, play a role, and not the failure rate in these intervals. Therefore, we might say that we postulate the following property of partial renewal: the mean number of failures in the interval I_k after a partial renewal carried out at the instant T_{k-1} equals the mean number of failures in the interval I_{k-1} multiplied by the degradation factor e^{α}.

We assume further that after K partial renewals the system undergoes a *complete renewal*. In the above references, this is termed an "overhaul." An overhaul brings the system failure rate to its initial level $h(0)$, i.e. to the "brand new" state.

Let us define the following costs: $c_{\min}$ is the cost of the minimal repair; c_{pr} is the partial renewal cost; C_{ov} is the overhaul cost. Simple reasoning leads to the following expression for the cost per unit time:

$$\eta(K) = \frac{c_{\min}H(T)(1 + e^{\alpha} + \ldots + e^{\alpha(K-1)}) + (K-1)c_{pr} + C_{ov}}{KT}. \qquad (4.2.8)$$

It is assumed that T is given. We are looking for an optimal K^* which minimizes $\eta(K)$. The above formula can be simplified by noting that $1+e^{\alpha}+\ldots+e^{\alpha(K-1)} = (e^{\alpha K} - 1)/(e^{\alpha} - 1)$.

Example 4.2.3
Suppose that $T = 1$, the failure rate $h(t) = 2t$, the degradation factor is $e^{\alpha} = 1.1$. The costs are: $c_{\min} = \$15$, $c_{pr} = \$150$, and $C_{ov} = \$1000$. Figure 4.7. shows the graph for $\eta(K)$ derived by *Mathematica*. It is seen that the optimal $K^* = 20$.

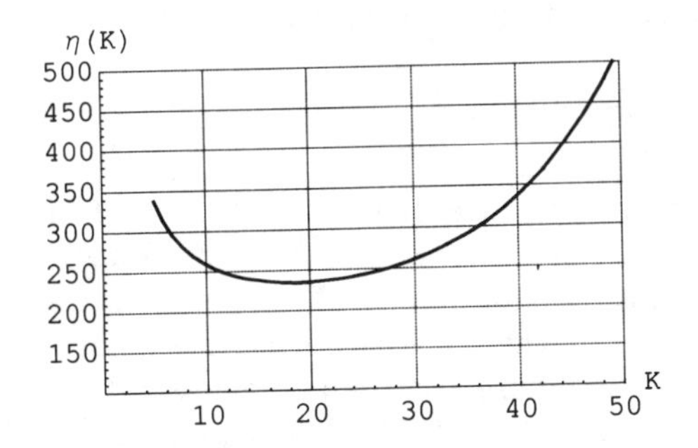

Figure 4.7. The graph of $\eta(K)$

Two quantities appear in (4.2.8) : $H = H(T)$ and α. For applications, it is vital to have an easy way of estimating these parameters from the data available. Suppose we register the numbers of failures N_i, $i = 1, \ldots, r$, in r consecutive intervals $I_1, I_2, \ldots, I_r$: $n_1, n_2, \ldots, n_r$. Theoretically, the mean number of failures M_i in I_i equals $M_i = e^{\alpha(i-1)}H$. Then

$$\log M_i = \log H + \alpha(i-1). \tag{4.2.9}$$

We suggest replacing $\log M_i$ by $\log n_i$ and using the linear regression technique for finding $a = \log H$ and α. In other words, these parameters must be found by minimizing the expression

$$S = \sum_{i=1}^{r} \Big(\log n_i - a - (i-1)\alpha \Big)^2. \tag{4.2.10}$$

So far we have considered *periodic* maintenance policies. These are convenient for implementation, but the choice of the period T has been made without taking into account the previous maintenance history, e.g. the actual *age* of the system. In the next subsection we consider the so-called age replacement policy, which does take into account the system age.

4.2.5 Age Replacement – Cost-type Criterion

At $t = 0$, a new unit starts operating. Its lifetime c.d.f. is $F(t)$. The unit is replaced when it reaches age T, or at its failure, whichever occurs first. The ER (repair upon failure) costs c_{ER}, the PM (preventive replacement at age T) costs $c_{PM} \ll c_{ER}$. Each replacement completely renews the system and takes a negligible time. Denote by τ the unit's lifetime. Let $Z = \min(\tau, T)$. Obviously, the first renewal appears after a random time Z. So the inter-renewal period has mean

$$E[Z] = \int_0^T (1 - F(x))dx;$$

see Example 1.3.2. The mean cost during one renewal period equals $F(T)c_{ER} + (1 - F(T))c_{PM}$. Therefore, the mean cost per unit time is

$$\eta_{age}(T) = \frac{F(T)c_{ER} + (1 - F(T))c_{PM}}{\int_0^T (1 - F(x))dx}. \tag{4.2.11}$$

Let us derive a formula for the mean time to failure in age replacement. A typical maintenance–repair history is as follows. A new component starts operating at $t = 0$. On a random number N of occasions the age T is reached before failure occurs, and on the $(N+1)$th cycle the failure takes place before age T is reached. Thus the failure appears at time $t_f = TN + \xi$, where $P(N = k) = (1 - F(T))^k F(T)$, $k \geq 1$, and $E[\xi]$ lies between 0 and T. It is easy to establish that $E[N] = (1 - F(T))/F(T)$ and thus the bounds on the mean time to failure are

$$\frac{T(1 - F(T))}{F(T)} < E[t_f] < \frac{T}{F(T)}. \tag{4.2.12}$$

Example 4.2.4: Age replacement of a chemical reactor
A chemical reactor producing cyanides has a system of pipes subject to the action of extremely active and dangerous chemicals. A post-failure repair of these pipes is very costly because of air pollution and contamination. The cost of ER is estimated as c_{ER} = \$100 000. PM of the pipes costs much less – c_{PM} = \$1000.

Expert opinion, based on past experience of using similar reactors, has it that the mean lifetime of a reactor is 1000 operation cycles. It is also assumed that the lifetime follows the Weibull distribution with shape parameter $\beta = 3$. Let us find the optimal replacement age T^*, the minimal costs $\eta_{age}(T^*)$ and the mean interval between reactor failures. Note that there is a safety constraint on the failure probability on the interval $[0, T]$: the probability of reactor failure should not exceed 0.01.

It follows from (2.3.12) that $\lambda = \Gamma[1 + 1/3]/1000 = 0.000\,893$. Thus $F(t) = 1 - \exp[-(0.000\,893t)^3]$, $c_{ER} = 100\,000$, $c_{PM} = 1000$.

Figure 4.8 shows the graph of $\eta_{age}(T)$. The optimal replacement age $T^* \approx 200$ cycles and $\eta_{age}(T^*) \approx$ \$8 per cycle. $F(200) = 1 - \exp[-(200 \times 0.000\,893)^3] = 0.0057$, which satisfies the safety constraint.

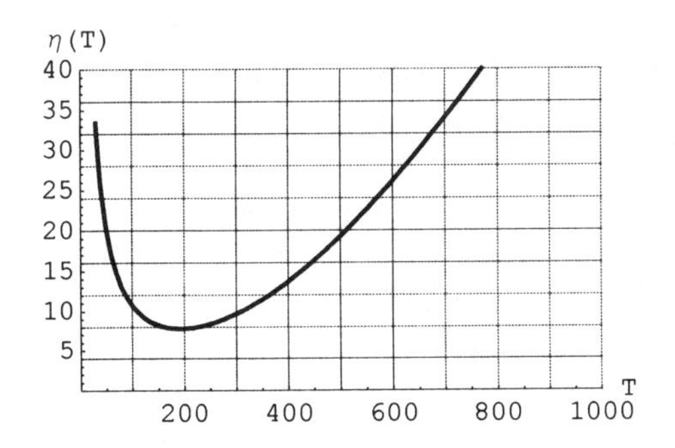

Figure 4.8. $\eta_{age}(T)$ in Example 4.2.4

It is interesting to compare this value with the cost of using the reactor if no PM is carried out. Then, the amount of \$100 000 is paid, on average, every 1000 cycles and thus $\eta_{age}(\infty) = 100$, which is about 12.5 times the value $\eta_{age}(T^*)$.

The mean interval between failures by (4.2.12) is approximately equal to $T^*/F(T^*) = 35\,200$ cycles .

4.2.6 Age Replacement – Availability-type Criterion

Let us preserve all the assumptions of Section 4.2.5 except that now PM and ER last t_{PM} and and t_{ER}, respectively. Typically, $t_{ER} \gg t_{PM}$. Denote by $f(t)$ the lifetime density.

One renewal period will be equal either to $X = t + t_{ER}$, if the unit fails in the interval $(t, t + dt)$, $t \le T$, which takes place with probability $f(t)dt$, or to $X = T + t_{PM}$, if the unit reaches age T, which takes place with probability $1 - F(T)$. Thus, after simple algebra,

$$E[X] = \int_0^T (1 - F(x))dx + t_{ER}F(T) + t_{PM}(1 - F(T)) \,. \tag{4.2.13}$$

In one renewal period the unit is operational, on average, for time $\int_0^T (1 - F(x))dx$. Suppose that the reward equals the operational time. Then the mean reward per unit time is the system stationary availability:

$$\eta_F(T) = \frac{\int_0^T (1 - F(x))dx}{\int_0^T (1 - F(x))dx + t_{ER}F(T) + t_{PM}(1 - F(T))}.$$

It is easy to simplify this expression:

$$\eta_F(T) = \left(1 + \frac{t_{ER}F(T) + t_{PM}(1 - F(T))}{\int_0^T (1 - F(x))dx}\right)^{-1}. \tag{4.2.14}$$

Thus we see that maximizing the availability is equivalent to minimizing the corresponding cost-type criterion (4.2.11).

4.3 Qualitative Investigation of Age and Block Preventive Maintenance

The basic expressions for the average costs are

$$\eta_{block}(T) = (c + m(T))/T,$$

for the block replacement and

$$\eta_{age}(T) = (F(T) + c(1 - F(T)))/\int_0^T (1 - F(x))dx,$$

for the age replacement.

4.3.1 The Principal Behavior of $\eta_{block}(T)$ and $\eta_{age}(T)$ as a Function of T – the Role of c and $F(t)$

Both expressions behave similarly when $T \to 0$: both tend to infinity. As $T \to \infty$, both expressions tend to $1/\mu$, where μ is the mean lifetime. We proved this fact earlier for block replacement. For age replacement, note that the denominator tends to μ as T tends to infinity and the numerator tends to 1.

There are two types of behavior of $\eta_{(\cdot)}(T)$, as shown in Fig. 4.9. Type I means that the minimal cost is attained at $T = \infty$, or the PM should be avoided. It easy to check that for $F(t) = 1 - e^{-\lambda t}$ this is the case, obviously for both age and block maintenance. Type II behavior means that there is a finite optimal maintenance period and/or optimal age T^*. If $c > 1$ or $F(t)$ has a DFR, we observe type I behavior.

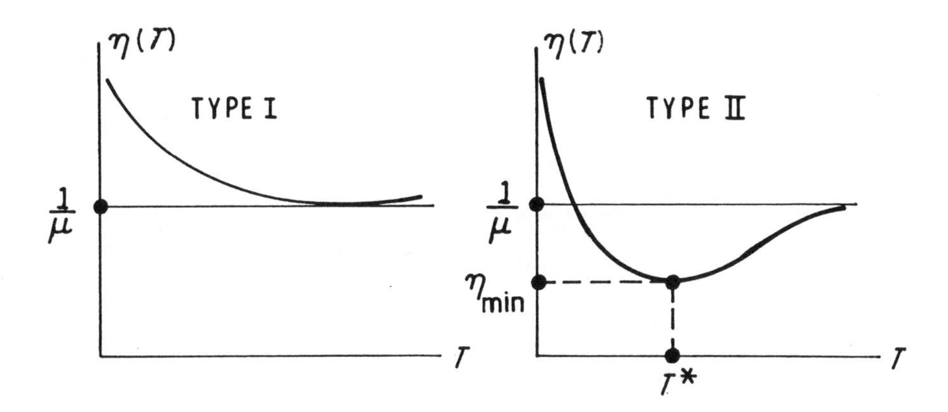

Figure 4.9. Type I and type II behavior

Figure 4.9 suggests a natural measure Q of preventive maintenance efficiency:

$$Q = \frac{\eta_{(\cdot)}(\infty)}{\min_{T>0} \eta_{(\cdot)}(T)} = \frac{1}{\eta_{(\cdot)}(T^*) \cdot \mu} . \tag{4.3.1}$$

For type I behavior, $Q = 1$; for type II behavior, $Q > 1$. The greater is Q, the more efficient is the optimal preventive maintenance.

As a general rule, we can say that Q increases with a decrease in c and with a decrease in the coefficient of variation of the lifetime $F(t)$. In simple words, preventive maintenance (with a proper choice of T^*) is more efficient if the PM costs less in comparison with the cost of ER, and if the density $f(t)$ is more "peaked" around the mean value.

Note that the largest possible value of Q will be attained for an "ideal" distribution: $F(t) = 0$ for $t < \mu$, and $F(t) = 1$ otherwise. For this case, the optimal $T = \mu - 0$ ("a little less" than μ), and the maximal $Q = 1/c$.

Table 4.1 shows the values of Q and T^* for various values of c and the coefficient of variation of $F(t)$. These calculations were done for block replacement, for the gamma distribution $Gamma(k, \lambda)$. The picture for age replacement is similar. Recall that $c.v. = 1/\sqrt{k}$.

It is seen from Tab. 4.1 that Q is nondecreasing for each c as the c.v. decreases. Similarly, for each fixed c.v., Q increases with the decrease in c. Typically, the optimal maintenance period T^* increases with a decrease in c.v. and an increase in c. It is also typical that the optimal T lies in the range $[0.4 - 0.65]\mu$.

Table 4.1: Q and T^* for block replacement

c	$k=2$	$k=4$	$k=8$	$k=16$	T^*, Q
0.15	0.42μ	0.4μ	0.47μ	0.56μ	T^*
	1.11	1.72	2.50	3.23	Q
0.35	–	0.57μ	0.58μ	0.64μ	T^*
	1	1.01	1.28	1.56	Q
0.50	–	–	–	0.68μ	T^*
	1	1	1	1.15	Q

Table 4.2: Dependence of Q_{age}/Q_{block} on c and k

k	$c=0.2$	$c=0.5$	k	$c=0.2$	$c=0.5$	k	$c=0.2$	$c=0.5$
2	1.10	1	4	1.08	1.03	8	1.05	1.14

4.3.2 Which is Better: Age or Block Replacement?

Suppose we apply age and block replacement to the same system, each of them in its optimal way. The PM in both schemes costs c; the cost of the ER is 1, as it is for age and block replacement. Let us compare $\min_{T>0} \eta_{age}(T)$ with $\min_{T>0} \eta_{block}(T)$.

It turns out that the optimal age replacement is more profitable than the optimal block replacement. To prove this, we need to involve the notions of optimal Markovian-type strategies; for details see Gertsbakh (1977, p. 99). Table 4.2 presents a comparison between age and block replacement for the gamma family and various c values.

This table shows that age replacement is always better than block replacement, but the ratio of the efficiencies Q_{age}/Q_{block} is relatively close to 1.

It should be noted that in practice, for the same ER cost, the cost of a PM carried out at a preplanned time, as in block replacement, will be *smaller* than the corresponding PM cost for age replacement.

4.3.3 Finding the Optimal Maintenance Period (Age) T^*

Formally, T^* must satisfy the equation

$$\frac{\partial \eta_{(\cdot)}(T)}{\partial T} = 0. \tag{4.3.2}$$

Simple formulas are not available for the root of this equation. It is a waste of time trying to find T^* analytically, and the problem should be tackled numerically. This is very simple using *Mathematica* software. Plot the corresponding

Table 4.3: Efficiency Q as a function of ϵ

β	$\epsilon = 0.0$	$\epsilon = 0.1$	$\epsilon = 0.2$	λ	$\epsilon = 0.0$	$\epsilon = 0.1$	$\epsilon = 0.2$	β
2	1.38	1.28	1.22	2	1.96	1.64	1.47	3
2	1.38	1.25	1.13	4	1.96	1.52	1.39	3
2	1.38	1.22	1.13	8	1.96	1.52	1.35	3

$\eta_{(\cdot)}(T)$ for $T < 2\mu$. Typically, if the optimal T exists, it lies within the interval $[0, 2\mu]$. Locate the optimal T and use the "FindMinimum" operator.

4.3.4 Contamination of $F(t)$ by Early Failures

In practice, the efficiency of the optimal preventive maintenance could be considerably impaired by contaminating the population $F(T)$ by an exponential population with much smaller mean value. Formally, let us assume that the lifetime distribution has the c.d.f.

$$F(t) = \epsilon F_1(t) + (1 - \epsilon)F_0(t), \tag{4.3.3}$$

where F_0 is the principal population with an increasing failure rate. It is contaminated with the addition of $100\epsilon\%$ of "bad" items having exponential lifetime $F_1(t) = 1 - e^{-\lambda t}$. Recall that for c.d.f. F_1, no optimal preventive maintenance policy exists. Let us consider a numerical example.

Example 4.3.1 : Optimal age replacement for a contaminated Weibull population Let us take $F_1(t) = 1 - e^{-\lambda t}$ and $F_0(t) = 1 - e^{-t^{\beta}}$, $\beta = 2$ and 3. Table 4.3 presents the results of computing Q for various ϵ and λ values, for $c = 0.2$. It is seen that even a small contamination of 10% may considerably reduce the efficiency Q.

4.3.5 Treating Uncertainty in Data

When we know exactly the lifetime distribution $F(t)$ and the maintenance cost c, the problem of finding the optimal maintenance period (or age) T^* is in principle a simple task, even if there are no closed-form expressions for T^*.

How do we handle the problem of finding the best maintenance policy if there is uncertainty with regard to F and c? To be more specific, let us consider an age replacement model with two possibilities regarding the c.d.f. $F(t)$, e.g. $F_1 \sim W(\lambda_1, \beta_1)$ and $F_2 \sim W(\lambda_2, \beta_2)$.

Suppose that these two distributions represent two extremes: the "best" and the "worst" case, respectively. We suggest defining these extremes according to the value of the coefficient of variation. So, F_1 has the smallest c.v. and F_2 has the largest c.v.

Suppose that the cost c lies between $c_{\min}$ and $c_{\max}$. Thus we consider four data combinations: data(1) $= (F_1,\ c_{\min})$; data(2) $= (F_1,\ c_{\max})$; data(3) $= (F_2,\ c_{\min})$; data(4) $= (F_2,\ c_{\max})$.

Figure 4.10 presents four curves of $\eta_{age}(T; data(i))$ for $i = 1, 2, 3, 4$. We wish to choose an appropriate age T for the PM. How do we do that in the presence of uncertainty?

One possible solution would be the "statistician's approach": let us assign a certain "probabilistic weight" to each data set and calculate the average cost $\hat{\eta}_{age}(T)$. Then we choose that value of T which minimizes $\hat{\eta}_{age}$. (The statistician would say that he/she is minimizing the ***average risk***.) A reasonable approach is to give equal weight to all four data combinations. Thus we write the expression

$$\hat{\eta}_{age}(T) = \frac{\eta_{age}(T; data(1)) + \ldots + \eta_{age}(T; data(4))}{4}, \tag{4.3.4}$$

and look for T^{**} which minimizes $\hat{\eta}$: $\min_{T>0} \hat{\eta}_{age}(T) = \hat{\eta}_{age}(T^{**})$. In the statistical literature $\hat{\eta}_{age}(T^{**})$ is called Bayesian risk (see DeGroot 1970, Sect. 8.2), and T^{**} is called the Bayesian decision. The Bayesian risk is shown in Fig. 4.10 as a dashed curve with open circles.

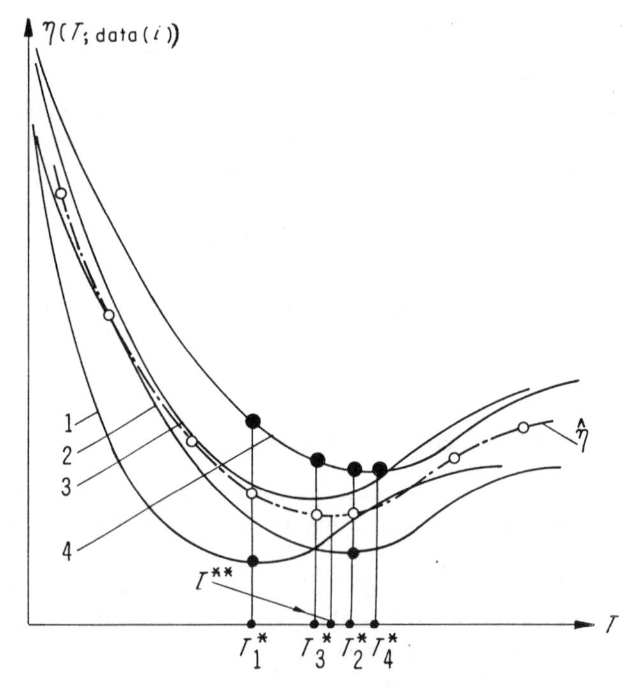

Figure 4.10. Curve i corresponds to $data(i), i = 1, 2, 3, 4$

Another approach, which I would call "the pessimist's solution," is the following: choose for each data set the best age T_i^*. Assume that "nature" plays against us and, as if by Murphy's Law, the real data would always correspond to the *maximal* cost for this age. For example, Fig. 4.10 shows that data set 4 is the worst for T_2^*. The maximal costs are shown by filled black dots, and the minimal costs by small filled dots. Now choose the data set which gives the *minimal* value of the maximal costs. For the curves in Fig. 4.10, this will be data(4). The corresponding "best" T is T_4^*.

Suppose that we have to choose between two extreme cases: the first is an IFR-type lifetime distribution with mean μ. The optimal maintenance age T^* guarantees $\eta_{age}(T^*) = 0.7/\mu$. The second is an exponential lifetime distribution with the same mean μ, for which, as we know, the optimal age is infinity and $\eta_{age}(\infty) = 1/\mu$. The "pessimist's approach" chooses in this situation the option "never do a PM," i.e. formally an infinite age for the replacement. This is consistent with the principle: "If it ain't broke, don't fix it."

Another version of the pessimist's approach is the so-called *minimax* principle. Rappoport (1998, p. 51) says "...that it rests on the assumption that the worst that can possibly happen will happen. Application of the principle amounts to expecting the worst possible outcome of each choice and choosing the alternative of which the worst outcome is the best of all the worst outcomes associated with the alternatives." See also DeGroot (1970, Sect. 8.7).

In our case, "the alternatives" are the replacement ages T. The implementation of the minimax principle to our situation means:
(i) calculating $\eta_{\max}(T) = \max\{\eta_{age}(T; \text{data}(1)), \ldots, \eta_{age}(T; \text{data}(4))\}$;
(ii) finding the value T_0 which would it minimize $\eta_{max}(T)$.

The minimax approach is illustrated in Fig. 4.11. The curve η_1 corresponds to an exponential distribution with mean $\mu_1 = 1$, and the curve η_2 to an IFR distribution with mean $\mu_2 = 1/2$. The dashed curve is $\eta_{max}(T)$. According to the minimax approach, we should choose the age T_0. The pessimist would choose T_2^*.

4.3.6 Age Replacement for the IFR Family

We have already mentioned in Sect. 2.2 that there are quite satisfactory bounds on $1 - F(t)$ if F is of IFR type and the first two moments are known. Let us use these bounds for finding the best age replacement policy. Recall that the standard expression for the average cost is

$$\eta_{age}(T) = \frac{F(T) + c(1 - F(T))}{\int_0^T (1 - F(x))dx},$$

or

$$\eta_{age}(T) = \frac{1 - (1 - c)(1 - F(T))}{\int_0^T (1 - F(x))dx}. \tag{4.3.5}$$

Let $[1 - F(t)]^*$ and $[1 - F(t)]_*$ be upper and lower bounds for $[1 - F(t)]$, respectively. Substitute the lower bound into (4.3.5). Then we obtain an upper bound on $\eta_{age}(T)$ which we denote $\phi_{\max}(T)$. Similarly, substitute the upper bound and obtain the lower bound denoted $\phi_{\min}(T)$. Therefore,

$$\phi_{\min}(T) < \frac{1-(1-c)(1-F(T))}{\int_0^T (1-F(x))dx} < \phi_{\max}(T) . \tag{4.3.6}$$

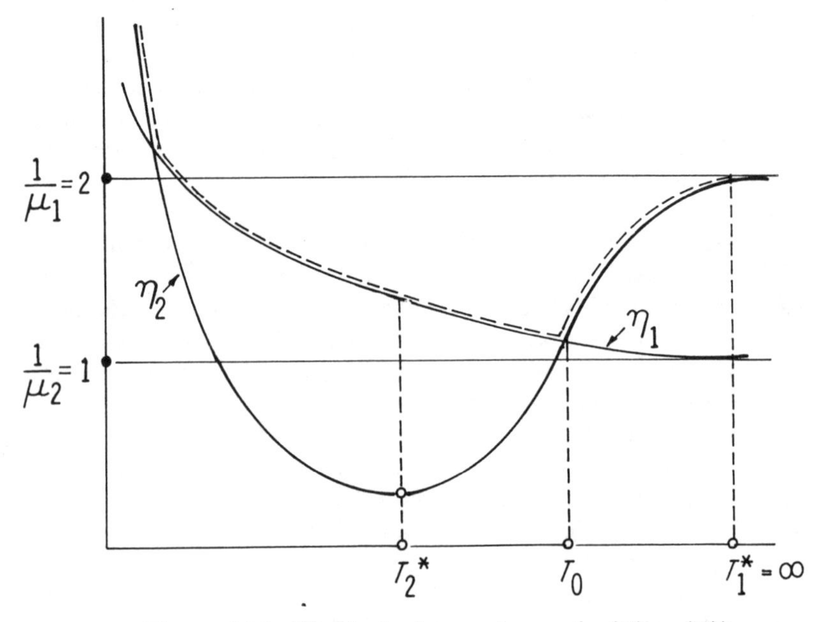

Figure 4.11. The dashed curve is $\max(\eta_1(T), \eta_2(T))$

Figure 4.12 shows the curves for $\phi_{\min}(T)$ and $\phi_{\max}(T)$, for $\sigma^2/\mu^2 = 0.2$ and $c = 0.1$. The coefficient of variation is $\sqrt{0.2} = 0.45$. The time scale has been changed to express the replacement age in units of the mean lifetime. This can always be achieved by replacing T by $T_1\mu$. If we choose $T_1^* = 0.5$, i.e. replace at age 0.5μ, our costs will be between $\phi_{\min} = 0.20$ and $\phi_{\max} = 0.65$. In the worst-case situation, the costs will not exceed 0.65. This means that in the worst case the efficiency of the age replacement is $Q = 1/\phi_{\max} = 1/0.65 = 1.54$, which is quite a satisfactory result.

More information on this topic can be found in Gertsbakh (1977, pp. 103-108).

4.3.7 Age Replacement with Cost Discounting

We have measured the efficiency of the preventive maintenance policy in terms of the mean costs *per unit time*. This was in fact a mathematical "trick" which helped us compare infinite costs over an infinite time period. But there are other ways to tackle infinity. Very often, the costs are calculated by means of so-called *discounting*. Each amount of money earned or spent at time t is converted to an equivalent amount at the actual starting time $t = 0$ by multiplying it by the discount factor $e^{-\alpha t}$, where α is a small positive number called the discount rate.

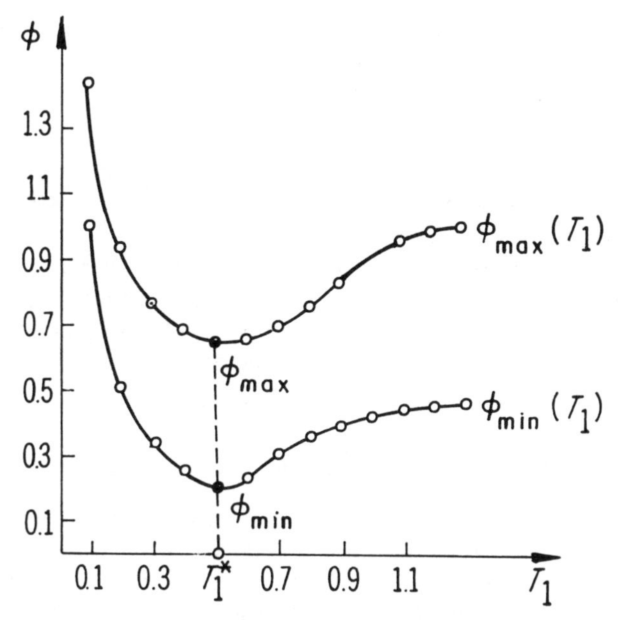

Figure 4.12. The upper and lower bound $\phi_{\max}(T_1)$ and $\phi_{\min}(T_1)$

Let us derive an expression for age replacement cost with discounting. Assume that PM costs c, EM costs 1, and the discount factor equals α. Denote by $C_\alpha(T)$ the discounted cost for replacement at age T. Then we can write the following recursive formula:

$$C_\alpha(T) = \int_0^T f(x)e^{-\alpha x}(1+C_\alpha(T))dx+(1-F(T))e^{-\alpha T}(c+C_\alpha(T)) \, . \tag{4.3.7}$$

Indeed, with probability $f(x)dx$ the failure takes place in $[x, x+dx]$, $0 \le x \le T$, and the cost of it is 1 plus the future cost $C_\alpha(T)$, all multiplied by the discount factor $e^{-\alpha x}$. If there is no failure in $[0,T]$, then PM is carried out at age T, and the corresponding cost $c + C_\alpha(T)$ is discounted by the factor $e^{-\alpha T}$.

If $\alpha \to 0$, the cost $C_\alpha(T)$ goes to infinity. The cost per unit time (without discounting) equals

$$C_0(T) = \frac{F(T) + c(1 - F(T))}{\int_0^T (1 - F(x))dx} \, . \tag{4.3.8}$$

The following theorem connects the discounted cost and the cost per unit time for *small α values.*

Theorem 4.3.1

$$\lim_{\alpha\to 0} \Big(\alpha C_\alpha(T)\Big) = C_0(T), \tag{4.3.9}$$

for all $T > 0$.

Proof

Express $C_\alpha(T)$ using (4.3.7) and multiply it by α:

$$\alpha C_\alpha(T) = \frac{(\int_0^T f(x)e^{-\alpha x}dx + c(1-F(T))e^{-\alpha T}) \cdot \alpha}{1 - \int_0^T f(x)e^{-\alpha x}dx - (1-F(T))e^{-\alpha T}} .$$

In this fraction, numerator and denominator tend to zero as α tends to zero. Use L'Hôpital's rule, differentiate the numerator and denominator with respect to α and set $\alpha = 0$. The result follows.

The practical conclusion from this theorem is that for small discount factor values (which is practically the case in reality), the cost per unit time $C_0(T)$ is approximately ***proportional*** to the discounted cost $C_\alpha(T)$. In practical terms this means that it does not matter which of these cost criteria are used for finding the best replacement age T^*.

4.3.8 Random Choice of Preventive Maintenance Periods

It was assumed in all the principal maintenance models considered in Section 4.2 that the maintenance period T or the maintenance age T is a ***nonrandom*** quantity which we choose in some "optimal" way to optimize costs or rewards. In practice, however, the maintenance period is always subject to certain random changes. For example, it has been decided to inspect the mechanical equipment of an elevator every half a year. The technicians who carry out the maintenance may start the inspection a week later (or a week earlier), depending on their workload in taking care of other elevators.

The "right" nomination of the inspection time for an elevator should be done in terms of operation hours. For example, it is assumed that the elevator works, on average, 8 hours daily. Suppose that the best inspection period is 1500 operation hours, which on the average is close to a half-year inspection period. But even if the elevator is inspected *exactly* every 182 days, the total amount of operating time between inspections will show considerable variations.

We have two tasks. First, we have to derive an expression for costs for the random maintenance period. Second, we have to investigate how the presence of randomness influences the expected costs. Suppose that we are able to ***control*** the randomness of the maintenance period, i.e. to choose the c.d.f. $G_T(t)$ of the maintenance period T. Is it possible to ***reduce*** the minimum expected costs by a proper choice of the c.d.f. $G_T(t)$? It is easy to derive an expression for costs under the condition of a random maintenance period. Assume that the random maintenance period T is independent on the system lifetime and has the density function $g_T(t)$. A typical expression for the average cost per unit time for a fixed maintenance period T has the form

$$\eta(T) = \frac{E[R(T)]}{E[L(T)]}, \tag{4.3.10}$$

where the denominator is the expected duration of the renewal period (which of course depends on T), and the numerator is the expected cost for one such renewal period. For example, in case of age replacement, $E[L(T)] = \int_0^T (1 - F(t))dt$, and $E[R(T)] = c_f F(T) + c_{pm}(1 - F(T))$, where $F(\cdot)$ is the c.d.f. of system lifetime, c_f and c_{pm} are the costs associated with the ER and PM, respectively.

Now, if T is a random variable itself, the expected duration of one renewal period becomes

$$L_0 = \int_0^\infty E[L(x)] g_T(x) dx. \tag{4.3.11}$$

Similarly, the expected cost for one renewal period is now

$$R_0 = \int_0^\infty E[R(x)] g_T(x) dx \,. \tag{4.3.12}$$

Then the expected cost per unit time becomes

$$\eta_0 = \frac{R_0}{L_0}. \tag{4.3.13}$$

For example, consider age replacement with T random, e.g. T is uniformly distributed in the interval $[t_{\min}, t_{\max}]$. Then

$$\eta_0^u = \frac{\int_{t_{\min}}^{t_{\max}} (c_f F(x) + c_{pm}(1 - F(x)))dx}{\int_{t_{\min}}^{t_{\max}} (\int_0^x (1 - F(t))dt}. \tag{4.3.14}$$

The following theorem demonstrates that no reduction in the expected costs can be obtained by introducing randomness in the maintenance parameter T.

Theorem 4.3.2
Let

$$\min_{T>0} \eta(T) = \eta(T^*) = \frac{E[R(T^*)]}{E[L(T^*)]} \,.$$

Then $\eta_0 \geq \eta(T^*)$. *Proof*

Since

$$\eta(x) = \frac{E[R(x)]}{E[L(x)]} \geq \frac{E[R(T^*)]}{E[L(T^*)]},$$

$$E[R(x)]E[L(T^*)] \geq E[R(T^*)]E[L(x)] \,. \tag{4.3.15}$$

Multiply both sides of (4.3.15) by $g_T(x)$ and integrate with respect to x from 0 to ∞. Then

$$\int_0^\infty E[R(x)]E[L(T^*)] g_T(x) dx \geq \int_0^\infty E[R(T^*)]E[L(x)] g_T(x) dx. \tag{4.3.16}$$

But (4.3.16) is equivalent to

$$\frac{E[R(T^*)]}{E[L(T^*)} \leq \frac{\int_0^\infty E[R(x)]g_T(x)dx}{\int_0^\infty E[L(x)]g_T(x)dx}, \tag{4.3.17}$$

which proves the theorem.

Example 4.3.2: Minimal repair with random maintenance period
Suppose that the system is maintained according to the minimal repair scheme; see Section 4.2.4. The failure rate is $h(t) = 0.5t$, the PM cost is $c = 0.2$, the failure cost is 1. Find the optimal period of minimal repair.

Consider a random choice of the minimal repair period T, $T \sim \text{Exp}(1)$, and compare the costs. In our case the cost criterion is $\eta_D(T) = (\int_0^T 0.5tdt + 0.2)/T = (0.25T^2 + 0.2)/T$.
The optimal period is $T^* = \sqrt{2c/a} = 0.89$. This corresponds to $\eta_D(T^*) \approx 0.45$.

The cost criterion for random T equals, according to (4.3.13),

$$\eta_0 = \frac{\int_0^\infty 0.25x^2e^{-x}dx + 0.2}{\int_0^\infty xe^{-x}dx}. \tag{4.3.18}$$

After a little algebra we obtain $\eta_0 = 0.52$. Interestingly, this is not much worse than the optimal result 0.45.

Remark
The expression (4.3.10) has a general form $\eta = \int f(x)dF(x)/\int g(x)dF(x)$, where $F(x)$ is the c.d.f. of some random variable. Very often, this c.d.f. is not known, and only partial information is available about it. For example, we might know only the first two moments (the mean and standard deviation) of the r.v. $X \sim F(x)$. An interesting mathematical theory developed by Stoikova and Kovalenko (1986) (see also Kovalenko et al (1997)) allows us to obtain bounds on η in the class of all c.d.f.s having the given values of the moments.

4.4 Optimal Opportunistic Preventive Maintenance of a Multielement System

4.4.1 Introduction

Preventive maintenance of real-life systems is characterized by three features: the maintenance policy involves many components (elements); most preventive maintenance actions usually involve several components simultaneously; the cost of a "joint" preventive action is considerably *smaller* than the total cost of similar maintenance actions if carried out *separately*. When several components are served simultaneously, then the instant chosen for maintenance may well not be "optimal" for all components involved. But nevertheless, there is a gain in cost and in convenience terms when a group of components undergoes simultaneous checking, inspection and maintenance. This type of servicing is termed *opportunistic*. Most complex real-life systems in fact have an "opportunistic" maintenance schedule, for example cars, passenger aircraft and the like. So far we have considered only one example of an opportunistic maintenance; see Example 4.1.3. Now we look at this in detail.

4.4.2 System Description – Coordination of Maintenance Actions – Cost Structure

We assume that the whole assembly under service has three levels: the element level; the subsystem level and the system level. It would be useful to follow the description by checking with the example in Fig. 4.13. The subsystems are denoted by indices $i = 1, 2$. The elements of subsystem i are denoted by (i, j), $j = 1, 2, \ldots$. Figure 4.13 shows that subsystem 1 has elements (1,1),(1,2), and subsystem 2 has elements (2,1),(2,2) and (2,3).

To plan and coordinate the maintenance actions, a common time axis t is introduced. It is convenient to consider t not as calendar time but as an *operational* time scale, for example the mileage scale. On this scale, the maintenance actions are carried out simultaneously for groups of elements. To simplify the maintenance scheduling and to allow joint maintenance actions to be carried out, a common time grid with a step Δ is introduced. One may assume that $\Delta = 1$. Any preventive maintenance action can be carried out only at times t^* which are multiples of Δ: $t^* = k\Delta$, where k is an integer.

Assume that the cost of a failure of element (i, j) is a_{ij}.

The cost of a preventive (i.e. scheduled) maintenance (PM) action of element (i, j) consists of *direct* costs c_{ij} plus set-up costs. The structure of these set-up costs is crucial in our problem. If an element (i, j) undergoes a PM, the cost paid *in addition* to c_{ij} is f_0 for involving the whole system *plus* f_i for involving the subsystem i to which this element belongs.

The "opportunistic" nature of the costs is reflected in the following. Suppose that at a certain instant $\Delta k = t$, a maintenance action is planned for elements which constitute a group G. Suppose that G involves a set S of subsystems. Then the total costs of servicing the whole group G will be

$$C(G) = \sum_{(ij)\in G} c_{ij} + \sum_{i\in S} f_i + f_0. \tag{4.4.1}$$

Don't be alarmed by this expression. In Fig. 4.13, at time $t = 6$, a PM is planned for elements $(1, 1)$, $(1, 2)$, $(2, 1)$, and (2,3). This is group G. The subsystems involved are 1 and 2; $S = \{1, 2\}$. The total PM cost will be c_{11} +

$c_{12} + c_{21} + c_{23} + f_1 + f_2 + f_0$.

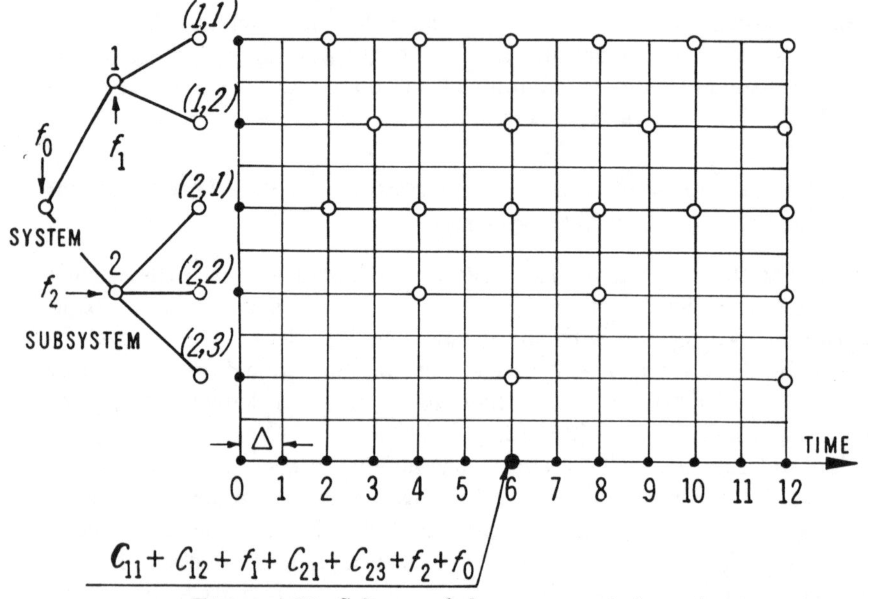

Figure 4.13. Scheme of the opportunistic maintenance

Let us consider an example.

Example 4.4.1

Write an expression of the costs per unit time for the system shown in Fig. 4.13. Let element (i, j) be serviced with period t_{ij}. Write $\mathbf{t} = (t_{11}, t_{12}, t_{21}, t_{22}, t_{23})$. Assume $t_{11} = 2, t_{12} = 3$; $t_{21} = 2, t_{22} = 4, t_{23} = 6$.

Each element is repaired according to the minimal repair scheme (see Section 4.2.4) and the average number of failures on $[0, t_{ij}]$ for element (i, j) is $H_{ij}(t_{ij})$.

The least common multiple (l.c.m.) of all t_{ij} values is 12 (set $\Delta = 1$). The whole preventive maintenance scheme is shown in Fig. 4.13. The mean costs per unit time are:

$$\begin{aligned}\eta(\mathbf{t}) &= \frac{\Big(6(H_{11}(2)a_{11} + c_{11}) + 4(H_{12}(3)a_{12} + c_{12})\Big)}{12} \\ &+ \frac{\Big(6(H_{21}(2)a_{21} + c_{21}) + 3(H_{22}(4)a_{22} + c_{22}) + 2(H_{23}(6)a_{23} + c_{23})\Big)}{12} \\ &+ \frac{8f_0 + 8f_1 + 6f_2}{12}.\end{aligned}$$

Let us explain this expression. The component (1,1) is repaired six times in the period (0,12], the component (1,2) four times, etc. This explains the coefficients of the $H_{ij}(\cdot)$ terms. Furthemore, subsystems 1 and 2 are involved in preventive maintenance 8 and 6 times, respectively. For example, subsystem 2 is "opened" at $t = 2, 4, 6, 8, 10, 12$. The total number of maintenance actions in the period $(0, 12]$ is eight. This explains the last term.

4.4.3 Optimization Algorithm. Basic Sequence

We wish to find an optimal vector $\mathbf{t}$ which would minimize $\eta(\mathbf{t})$. A formal description of the optimization algorithm needs requires the notion of a ***basic sequence***. Suppose that for a subsystem i we have a set of preventive maintenance periods $\mathbf{t}_0 = (t_{i1}, \ldots, t_{im})$. Without loss of generality, we may assume that this sequence is ordered, so that $t_{i1} \leq \ldots \leq t_{im}$.

By definition, the basic sequence consists of vectors $\mathbf{t}_0, \mathbf{t}_1, \mathbf{t}_2, \ldots$, where each successive vector is obtained by increasing by Δ the ***minimal*** component(s) of the previous one. For example, if $\mathbf{t}_0 = (2, 4, 5)$ then $\mathbf{t}_1 = (3, 4, 5)$; $\mathbf{t}_2 = (4, 4, 5)$, $\mathbf{t}_3 = (5, 5, 5)$, etc.

The optimization algorithm consists of the following steps:

Step 1: Individual (element) optimization. At this step, an optimal PM period is found for each element separately, without taking into account the subsystems and system set-up costs.

Step 2: Optimization at subsystem level. Take the optimal periods of PM found for elements of a subsystem i in step 1 and organize them into a vector $\mathbf{t}_i$. This will be the initial vector of the basic sequence for this subsystem. The basic sequence for subsystem i is obtained according to the above description. For each vector of this basic sequence $\mathbf{t}_i^0, \mathbf{t}_i^1, \mathbf{t}_i^2, \ldots$, calculate the costs per unit time for the ith subsystem. These costs include the set-up costs f_i only.

Denote by $\mathbf{t}_i^*$ the best vector for subsystem i.

Step 3: Optimization at system level. Concatenate all subsystem optimal vectors $\{\mathbf{t}_i^*\}$ obtained in step 2 into one vector for the whole system. Denote it by $\mathbf{t}_0^0$. Construct the corresponding basic sequence. Optimize the total costs per unit time on the vectors of this sequence. Take into account all set-up costs, including f_0. The vector from the "all-system" basic sequence $\mathbf{t}_0^*$ which minimizes the costs is the optimal PM vector.

Example 4.4.1 continued

Let us implement the above algorithm for the system of Fig. 4.13, for the following data: all $a_{ij} = 1$. $c_{11} = 0.1, c_{12} = 0.4$; $c_{21} = c_{22} = c_{23} = 0.2$. Also, $H_{11}(t) = t^2$; $H_{12}(t) = 0.25t^2$; $H_{21} = t + t^2$; $H_{22}(t) = 0.1t^2$; $H_{23}(t) = 0.05t^2$. The set-up costs are $f_1 = f_2 = 5$, $f_0 = 8$.

Step 1: Optimization at element level.

For element (i, j), the expression for minimization is

$$A_{ij}(t) = \frac{H_{ij}(t)a_{ij} + c_{ij}}{t}. \tag{4.4.2}$$

Simple calculations show that the optimal periods for elements (i,j) are: $t^*_{11} = 1$, $t^*_{12} = 1$; $t^*_{21} = 1$, $t^*_{22} = 2$, $t^*_{23} = 2$.

Step 2: Minimization at subsystem level.

Subsystem 1

The basic sequence is $(1,1),(2,2),(3,3),\ldots$. We have to minimize the sum of individual costs plus the contribution of the subsystem 1 set-up costs.

Let d_1 be the l.c.m. of t_{11}, t_{12} and let $M_1(t_{11}, t_{12})$ be the number of PMs on the interval $(0, d_1]$. Then add the subsystem 1 set-up cost $M_1 f_1/d_1$. For example, if $t_{11} = 2, t_{12} = 3$, then $d_1 = 6$ and $M_1 = 4$, since the PMs are done at the instants $2,3,4,6$. Then one should add $4f_1/6$.

For subsystem 1 we have to minimize the following expression:

$$B_{x,y} = \frac{x^2 + 0.1}{x} + \frac{0.25y^2 + 0.4}{y} + \frac{5}{x}\,.$$

(In our example $x = y$; the subsystem 1 is "opened" every x units of time). For $(1,1)$ we have $B_{11} = 6.75$. For $(2,2)$, $B_{2,2} = 5.25$. For $(3,3)$ we obtain $B_{3,3} = 5.58$, etc. The optimal vector is (2,2).

Subsystem 2

For subsystem 2, the basic sequence is (1,2,2), (2,2,2), (3,3,3),.... Similarly to subsystem 1, we obtain that the best vector is (2,2,2).

Step 3: The system level. The basic sequence for the whole system is $(2,2,2,2,2),(3,3,3,3,3),(4,4,4,4,4),\ldots$. Use for minimization the following principal expression:

$$\eta(x_{11},\ldots,x_{23}) = \sum_{i,j} \frac{H_{ij}(x_{ij})a_{ij} + c_{ij}}{x_{ij}} + \frac{M_1 f_1}{d_1} + \frac{M_2 f_2}{d_2} + \frac{M_0 f_0}{d_0}. \quad (4.4.3)$$

M_i and d_i require explanation. d_i is the least common multiple of $(x_{i1}, x_{i2}, \ldots)$. For example, if $x_{i1} = 2, x_{i2} = 3, x_{i3} = 4$, then $d_i = 12$. M_i is the total number of PMs made to subsystem i on $(0, d_i]$. For the above numbers, $M_i = 8$. d_0 is the l.c.m. of $(x_{11},\ldots,x_{23})$, and M_0 is the total number of PMs on $(0, d_0]$.

In our example, the calculations are very easy because all x_{ij} are the same for each vector of the basic sequence. The minimal costs are obtained for (3,3,3,3,3) and are equal to 14.6.

The main advantage of the above-described algorithm is that one must investigate only the vectors of the basic sequence. This algorithm provides the exact minimum if

(i) the time intervals allowed for PM form a sequence of type Δ, Δk_1, $\Delta k_1 k_2, \ldots$, where $k_1, k_2, \ldots$ are integers. For example, the following sequence satisfies this property : 1, 2, 4, 12 , 24,...;

(ii) $H_{ij}(x)$ are convex functions of x.

(ii) is satisfied if the component lifetimes are of IFR type.

We believe that in the general case our algorithm provides a solution close to optimal. More details about this algorithm can be found in Gertsbakh (1977, pp. 216–227).

4.4.4 Discussion – Possible Generalizations

1. The additive structure of PM costs, as described above, might be adequate for servicing a truck. The cost f_0 reflects the cost of removing the truck from service and of taking it to the service station; f_1 is the cost paid every time the engine is inspected, diagnosed and repaired; f_2 is the payment for inspecting and testing the braking system, etc. It is a reasonable assumption that all these costs must be added together.

A quite different situation is servicing a military aircraft. Here the main purpose is to achieve the maximal readiness (availability). Instead of costs it would be wise to look at the ***duration*** of servicing aircraft systems, i.e. the time it takes to service these systems. If the fuel system and engines are served simultaneously by two different service teams, then the true "cost" would be the service duration determined by the longest service time.

2. Optimal planning of maintenance activities has a lot in common with scheduling problems. The following example, suggested by George et al (1979), is a good illustration of how replacement and "classical" scheduling are related to each other. Suppose we have to maintain in proper condition the engines of a two-engine aircraft. Each engine has a limited resource in terms of hours of continuous operation. The following stock of engines is available: one engine with 100 hrs resource; four engines with 300 hrs resource; one engine with 200 hrs resource; one engine with 500 hrs resource. The total resource of all engines is 2000 hrs. A schedule which allows operation of the aircraft for 1000 flight hours is shown in Fig. 4.14a.

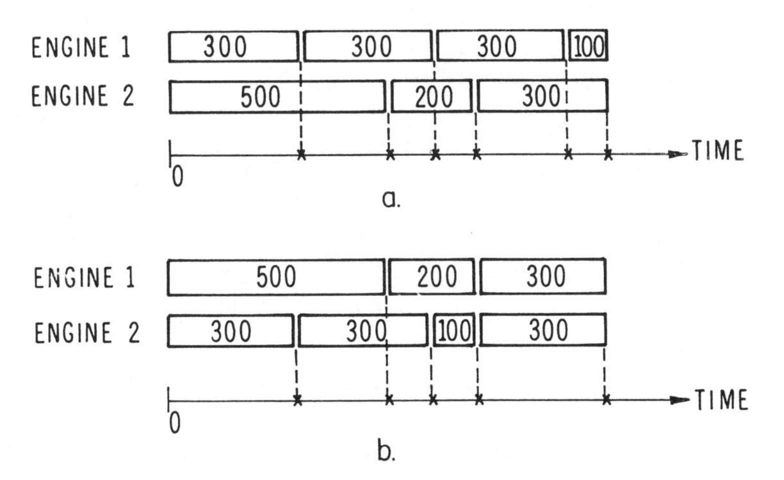

Figure 4.14. Two engine replacement schedules

The replacement of one engine or of two engines ***simultaneously*** costs $c = 1$. Thus the cost of the schedule in Fig. 4.14a is 6.

Clearly, it is not the most economic solution. A relatively small enumeration reveals that there is a cheaper schedule which is shown in Fig. 4.14b. Here the total cost of engine replacements is 5.

Finding an optimal replacement schedule for a fleet of several aircraft may be quite a challenging combinatorial problem.

The crucial feature of engine replacement schedule is that the simultaneous replacement of two engines costs half as much as two individual replacements. Cost reduction due to the simultaneous performance of several maintenance actions is typical of any opportunistic replacement scheme.

4.5 Exercises

1. You have at your disposal the following data on failures of the compressor unit of a new type of industrial refrigerator: 127 units started operating on January 1, 1990; 8 of them failed during the first two ears of operation; a total of 58 failures was registered during the period [1990, 1994], i.e. from January 1, 1990 to December 31, 1994.
a. Assume that the compressor's lifetime has a lognormal distribution. Equate $t_1 = 2$ years to the $8/127$ quantile, and $t_2 = 5$ years to the $58/127$ quantile. Estimate the parameters.
b. Assume that each failure of the compressor costs \$3500, and each preventive replacement costs \$1000. Would you advise carrying out an age replacement at age $T = 3$ years?

2. A machine for cotton production has two identical, independently failing mechanical parts. Two service strategies are suggested:
(i) replace each part upon its failure only;
(ii) at any failure, replace also the nonfailed part.

Each replacement completely renews the corresponding part. An individual replacement costs \$1500. Replacing the nonfailed part together with the failed part costs \$2000. There is uncertainty regarding the lifetime distribution of the parts. The first hypothesis is that $\tau \sim Gamma(k = 7, \lambda = 1)$. The second hypothesis is that $\tau \sim Gamma(k = 4, \lambda = 0.5)$.

Analyze replacement strategies (i) and (ii) under both hypotheses and make your suggestions regarding the "best" maintenance policy.

3. An age replacement is considered for bearing unit of a paper production line. The lifetime (in years) is $\tau \sim W(\lambda = 1, \beta = 4)$. The cost of emergency repair (upon failure) lies within the limits $c_e = [5, 20]$. The cost of the PM is $c_p = 1$. Make a recommendation regarding the optimal age of the preventive maintenance.

4. A street lamp is replaced periodically, with the replacement period T being

a random variable $T \sim N(\mu = 1, \sigma^2 = 0.1)$. The lifetime of the lamp is $\tau \sim Gamma(k = 4, \lambda = 1)$. Each failure costs \$100, each preventive maintenance costs \$5. Derive an expression for costs per unit time and evaluate it numerically.
Hint: Consider first a fixed period T. The costs per unit time have the form $\eta(T) = A(T)/T$. Argue that for random T, one should take $\eta = E[A(T)]/E[T]$, where the mean is taken with respect to the distribution of T.

5. The lifetime of a mechanical part has a Weibull distribution with mean 2000 hrs, and $c.v. = 0.3$. It has been decided to carry out an age replacement of the part. The ER takes 50 hrs and completely renews the part. The PM (which also renews the part) lasts 10 hrs. Find the optimal age T^* for the age replacement which would maximize the stationary availability.
Hint: Derive an expression for the stationary availability and reduce the problem to finding an optimal age replacement.

6. A system consists of two parts denoted 1 and 2. Each part undergoes a periodic repair. The repair period can be $T = \Delta,\ 2\Delta,\ 4\Delta,\ 8\Delta,\ 16\Delta$ (set $\Delta = 1$). Failures between scheduled repairs are eliminated according to the minimal repair scheme. The failure rates of the parts are: $\lambda_1(t) = 0.1t$; $\lambda_2(t) = 0.06t^2$. Each failure costs \$10, each scheduled repair costs \$2. The system set-up cost paid at each scheduled repair is $f_0 = 8$. Find an optimal opportunistic replacement schedule for the system.

7. Two-new type heavy trucks were monitored for 50 000 and 70 000 miles, respectively. Engine failures were observed at 26 000, 32 500 and 43 300 miles in the first truck, and at 19 500, 35 300, 50 200, and 68 700 miles in the second truck.
a. Assume that engine failures occur according to a nonhomogeneous Poisson process (NHPP) with intensity function $\lambda(t) = \exp[\alpha + \beta t]$. Estimate α and β using the maximum likelihood method described in Appendix A.
b. Each failure costs, on average, \$1000 and each preventive repair costs between \$100 and \$300. Assume that the engines are serviced according to the minimal repair scheme (Section 4.2.4). Assume that the engine failures appear according to a NHPP with intensity function $\lambda(t) = \exp[\hat{\alpha} + \hat{\beta} t]$, where $\hat{\alpha}$ and $\hat{\beta}$ are the MLEs found in part **(a)**. Make a recommendation regarding the optimal minimal repair period.

8. *Optimal periodic inspection of chemical defense equipment.*
Chemical defense equipment (CDE) is periodically inspected during its storage

in order to maintain its readiness. During the inspection, checks are made on the capability of the CDE to neutralize active ingredients of chemical weapons in the case of a chemical attack.

The interval between inspections is T, and the inspection lasts time t_0. The time t_0. The storage–inspection scheme is shown in Fig. 4.15. At the instant $t = 0$, the CDE is put in storage. During the interval $[T, T + t_0]$, the CDE is inspected. If the CDE has failed in the interval $[0, T]$, then the inspection reveals the failure, and the CDE is completely renewed toward the end of the inspection period. If there is no failure, the next inspection starts at the instant $2T + t_0$, and lasts t_0. It will reveal the failure if it appeared after the last inspection, and so on.

Assume that the lifetime of the CDE has a known density function $f(t)$.

The CDE is available for use ("ready") if it has not failed, and only between inspections. (It is not available during the inspections.) It is assumed that the "lifetime failure clock" works during the inspection and between the inspections in the same way.

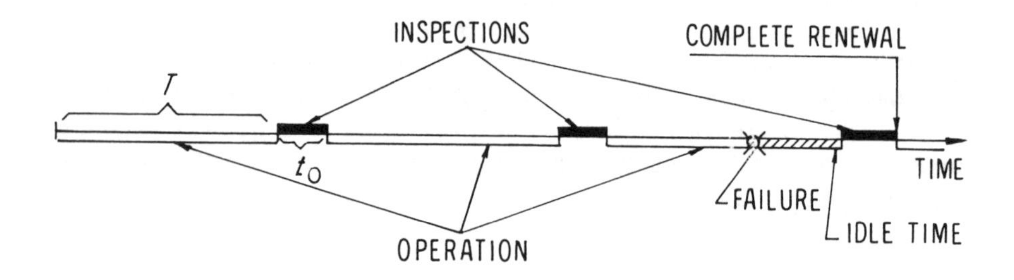

Figure 4.15. Operation–inspection scheme of the CDE

It will be assumed that if the CDE fails during the inspection, this fact will be immediately revealed, and the CDE will be completely renewed toward the end of the inspection.

Our task is to find the optimal value T^* of the inspection period T to ***maximize*** the ***stationary availability*** of the CDE. Denote by $\eta(T)$ the stationary availability of the CDE. Investigate $\eta(T)$ numerically for $f(t) = e^{-t}$ and t_0 in the range 0.05–0.1.

Hint: $\eta(T) = A(T)/B(T)$, where $A(T)$ is the "ready" time of the CDE during one storage–inspection–renewal cycle, and $B(T)$ is the mean duration of this cycle.

The expression for $A(T)$.

Suppose that the CDE fails at the instant x, which lies inside the "ready" period: $x \in [k(T+t_0), k(T+t_0)+T]$, where $k = 0, 1, 2, \ldots$. Then with probability $f(x)dx$, the "ready" time equals $x - kt_0$.

If the CDE fails during the inspection, i.e. $x \in [k(T+t_0)+T, (k+1)(T+t_0)]$, then the "ready" time equals $(k+1)T$. This leads to the following expression:

$$A(T) = \sum_{k=0}^{\infty} \int_{a_k}^{a_k+T} (x - kt_0) f(x) dx + \sum_{k=0}^{\infty} [F(a_{k+1}) - F(a_{k+1} - t_0)](k+1)T,$$

where $a_k = k(T + t_0)$ and $F(t) = \int_0^t f(x)dx$.
The expression for B(T).
If the failure appears in the interval $[a_k, a_{k+1}]$, the cycle length equals $(k+1)(T + t_0)$. This leads to the expression

$$B(T) = \sum_{k=0}^{\infty} (k+1)(T+t_0)[F(a_{k+1}) - F(a_k)] \,.$$

9. The following problem is discussed by Elsayed (1996, p. 530).

A critical component of a complex system fails when its failure mechanism enters one of two stages. Stage 1 is entered with probability α and stage 2 with probability $1-\alpha$. The lifetime of the component in stage i is exponentially distributed with parameter λ_i, $i = 1, 2$. Determine the optimal preventive maintenance interval for different values of λ_1 and λ_2, if the failure costs $c_f = \$500$ and the preventive maintenance costs $c_p = \$100$. (The author meant a block replacement.) What is, in your opinion, the optimal period of the preventive maintenance?

10. *Age replacement with a minimum-type exponential contamination.*
Consider an age replacement applied to a series system consisting of two independent components. The lifetime of the first component is $\tau_1 \sim \text{Exp}(\lambda_1)$, and the lifetime of the second component is $\tau_2 \sim W(\lambda_2 = 1, \beta = 2)$. The failure costs $c_f = 1$, the preventive maintenance costs $c_p = 0.2$.

Find the value of λ_1 which would give $E[\tau_1] = E[\tau_2]$. Write the expression for the cost functional $\eta(T)$, where T is the replacement age. Investigate $\eta(T)$ numerically and graphically and find the optimal T value and the corresponding minimal cost. Compute also the efficiency Q of the optimal age replacement.

Repeat the calculations for a system which has only one (the second) component. You will obtain much higher efficiency. Explain.

11. *Comparison of optimal age replacement policies.*
Compare $\eta_{age}(T)$ for two distributions $logN(\mu, \sigma)$ and $W(\lambda, \beta)$, which have the same mean 1 and the same coefficient of variation 0.25. Assume that the failure

costs $c_f = 1$, and that the PM costs $c = 0.2$.

12. ***Minimal repair with partial renewal.***
A machine is repaired according to the minimal repair scheme with partial renewal (Section 4.2.4). It is repaired at the instants $T_0 k$, $k = 1, 2, 3$. The degradation factor equals e^{α}. The machine was observed on the first five consecutive intervals and the following numbers of failures were recorded: $n_1 = 7, n_2 = 9, n_3 = 12, n_4 = 16, n_5 = 21$. The interval between incomplete renewals is $T_0 = 1$. Estimate the mean number of failures $H(1) = \int_0^1 h(v)dv$ and the degradation factor e^{α}. Assume the following costs: $c_{\min} = \$15$, $c_{pr} = \$150$, and $C_{ov} = \$1000$. Find the optimal period K^* for equipment overhaul.

Chapter 5

Preventive Maintenance Based on Parameter Control

Everywhere in life, the question is not what we gain, but what we do
Carlyle, *Essays*

5.1 Introduction – System Parameter Control

So far we have distinguished two system (component) states – up and down, or failure and nonfailure. The statistical information about the system (component) was related to its *lifetime*, i.e. to the transition time from the nonfailure ("good") state into the failure ("bad") state. We can picture the state of the system as a binary random process, see Fig. 5.1a. In this chapter we consider several models of preventive maintenance for which we know enough to distinguish *intermediate* states between the "completely new" and the "completely failed" system. A good example might be the description of failure by means of a damage accumulation model – recall the model leading to the gamma distribution. Preventive maintenance of a system with many states is an intervention in the damage accumulation process ξ_t by "pulling it down" from a dangerous state into the "safe" state, see Fig. 5.1b.

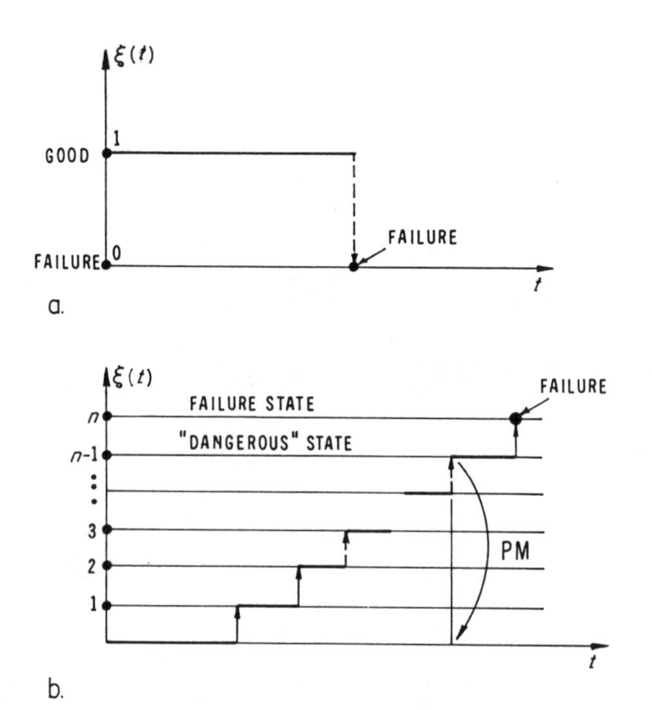

Figure 5.1. (a) A binary system. (b) Shifting system parameter from state $n-1$ to state 0

Mechanical deterioration is a good example of a process in which we can distinguish intermediate states between "brand new" and failed . The recent paper by Redmont et al (1997) studies the deterioration of concrete constructions (e.g. bridges) and defines 12 stages of failure development, from a hairline crack to large-volume splitting and corrosion. A system with n standby units allows a natural definition of intermediate states $\xi(t) = i,\ i = 0, 1, 2, \ldots, n$, according to the number i of failed standby units.

The most advanced preventive maintenance methods are based on a periodic or continuous follow-up of system diagnostic parameters, such as vibration monitoring or output fluid control; see Williams et al (1998). We discuss in Sect. 5.7 the preventive maintenance strategy based on a following up a multidimensional prognostic parameter.

Obviously, one might expect that preventive maintenance based on a periodic or continuous monitoring of system diagnostic parameters must be more efficient than a preventive maintenance policy based only on the knowledge of the existence of "good" and "bad" states. There is, however, a price for this higher efficiency: we have to know the statistical properties of the random process $\xi(t)$ which represents the damage accumulation and/or serves as a prognostic parameter for system failure. Usually this is a more involved task than just describing in statistical terms the transition time from $\xi(t) =$"brand new" into $\xi(t) =$"failure".

5.2 Optimal Maintenance of a Multiline System

We consider in this section a system consisting of n identical subsystems (lines). Every line operates and fails independently of the others, while the replacement (repair) of the failed line requires that the work of all lines be stopped. During the operation of the system, information is received about lines' failures. Define the state of the system $\xi(t)$ as the number of failed lines at time t. We assume therefore that we observe $\xi(t)$. We will also assume that the failure-free operation time of every line has a known c.d.f. $F(t)$.

System maintenance is organized as follows. When the number of failed lines $\xi(t)$ reaches the prescribed number k, $0 < k \leq n$, the system is stopped for repair, all failed lines are repaired, and all nonfailed lines are rendered "brand new". Thus the maintenance policy means shifting $\xi(t)$ from "dangerous" state $\xi = k$ into the initial state $\xi = 0$.

The real-life realization of the multiline system are certain types of mass production lines, and it is natural to assume that the reward from them is proportional to the total failure-free operation time. So we assume that each line gives a unit reward during a unit of failure-free operation. Denote by $t(k)$ the time needed to repair the system. Our purpose is to find the optimal k which provides the maximal mean reward per unit time.

Denote by τ_i the failure-free operation time of line i, $i = 1, 2, \ldots, n$. Denote by $\tau_{(i)}$ the ith ordered failure instant. The total system operation time during one operation-repair cycle is then

$$W(k) = n\tau_{(1)} + (n-1)(\tau_{(2)} - \tau_{(1)}) + \ldots + (n-k+1)(\tau_{(k)} - \tau_{(k-1)}). \quad (5.2.1)$$

The length of one operation-repair cycle is

$$L(k) = \tau_{(k)} + t(k). \quad (5.2.2)$$

Recalling the theory from Section 4.1, we arrive at the expression for the mean reward per unit time:

$$v(k) = \frac{E[W(k)]}{E[L(k)]}. \quad (5.2.3)$$

In order to be able to compare maintenance efficiency for systems with different n values, it is convenient to characterize the efficiency of the maintenance rule by the ratio

$$\chi_k = v(k)/n. \quad (5.2.4)$$

We assume that each line has a constant failure rate λ.

The expression for $W(k)$ is in fact identical to the formula for the total time on test; see Exercise 4, Section 2.4. Thus $W(k) \sim Gamma(k, \lambda)$ and

$$E[W(k)] = k/\lambda. \tag{5.2.5}$$

For $\tau \sim \text{Exp}(\lambda)$, $E[\tau_{(k)}] = \lambda^{-1} \sum_{i=1}^{k} (n - i + 1)^{-1}$. Using simple algebra we obtain that

$$\chi_k = \frac{1}{n} \frac{k}{\sum_{i=1}^{k} (n - i + 1)^{-1} + t(k)\lambda}. \tag{5.2.6}$$

Suppose that the repair time of k lines is

$$t(k) = N + Lk. \tag{5.2.7}$$

Consider first the case $N = 0$. It is easy to show that the denominator of χ_k increases with k, and therefore the optimal $k = 1$. In other words, for $N = 0$, the best policy is to stop the system for the repair immediately after the first failure of a line.

Now assume that the repair time is constant and does not depend on k, i.e. $t(k) = N$. Then the optimal k is usually not equal to one.

Figure 5.2 shows several graphs for the reward per unit time, for various values of $n = 4, 6, 8, 12$ and $N = 0.1, 0.2, 0.3$. (We assume that $\lambda = 1$). It will be seen that an incorrect choice of k may considerably reduce the reward. For example, for $n = 8$ and $N = 0.2$, the best $k = 3$. It provides a 50% increase in χ in comparison with a wrong choice of k, say $k = 1$.

Remark

The problem considered in this section is very similar to the block replacement of a group of machines; see Section 4.2.3. The maintenance policy considered there does the following: all machines are renewed after time T has elapsed since the last repair. Thus, the decision was based on *time*. A similar problem was considered by Okumoto and Elsayed (1983). In this section, the decision to renew the machines is based on observing the number of failed machines ("lines"). An interesting theoretical question is which policy can provide better results. It turns out that the answer is not trivial, and involves some advanced theoretical considerations. So, if the machines have identically distributed exponential lives, the optimal maintenance policy among all possible policies belongs to the subclass of the so-called Markovian policies. To put it simply, these are the maintenance rules which are based on observing the process $\xi(t)$. The best maintenance decision belongs to the class "repair after $\xi(t)$ has reached some critical level". This issue is discussed fuller in Gertsbakh (1984).

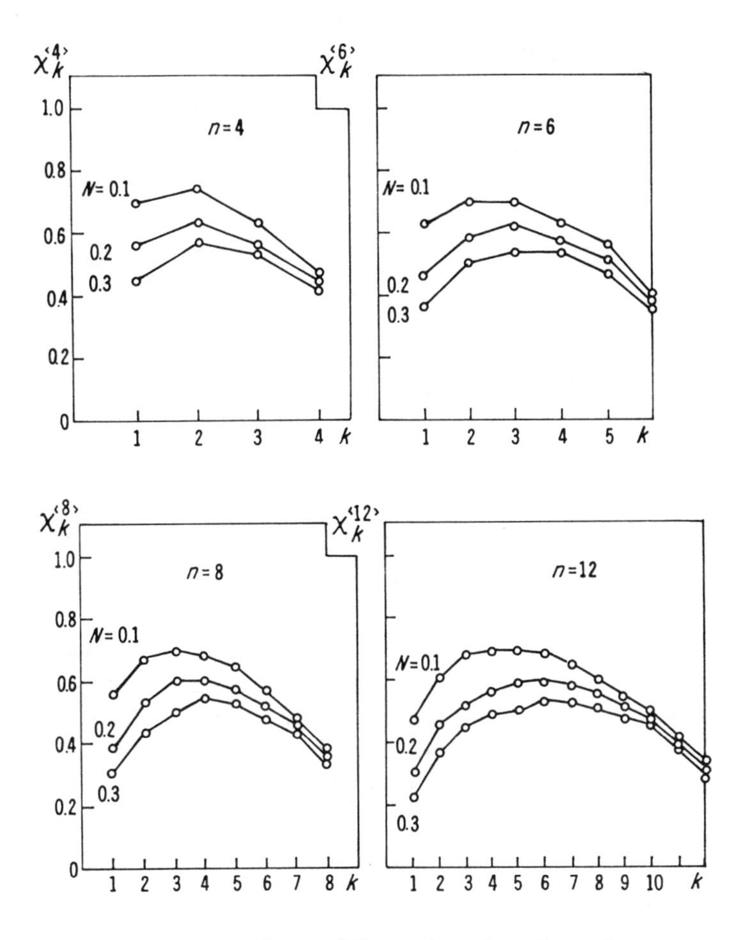

Figure 5.2. Reward for various k and n values

5.3 Preventive Maintenance in a Two-Stage Failure Model

5.3.1 The Failure Model

We distinguish *initial* failures and *terminal* failures. The initial failures appear according to a nonhomogeneous Poisson process (NHPP) with rate $\lambda(t)$. This means that an initial failure appears in the interval $[t, t+\Delta]$ with probability $\lambda(t)\Delta + o(\Delta)$ as $\Delta \to 0$.

Each initial failure *degenerates* into a *terminal* failure after a random time X; see Fig. 5.3. An initial failure which appears at time τ_i degenerates into a terminal failure after time X_i, i.e. the terminal failure appears at the instant $Z_i = \tau_i + X_i$. We assume that $\{X_i\}$ are i.i.d., independent on τ_i, and that $X_i \sim F(t)$.

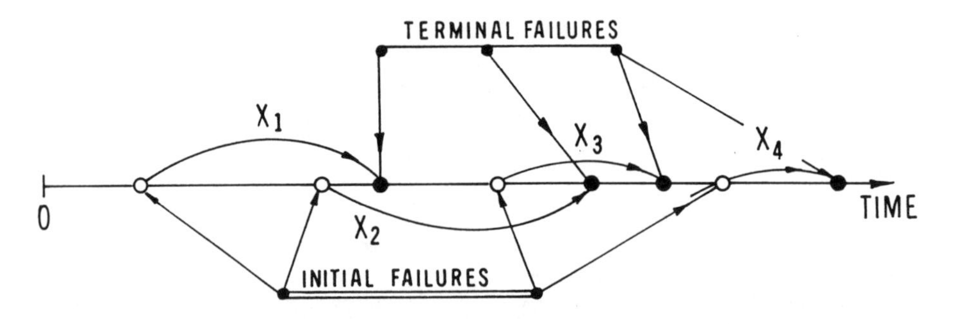

Figure 5.3. Degeneration of initial failures into terminal failures

5.3.2 Preventive Maintenance. Costs

The whole process of failure appearance and preventive maintenance is considered on a finite time interval $[0, T]$; see Fig. 5.4. Preventive maintenance actions are scheduled at the instants $t_1, t_2, \ldots, t_k$, $0 < t_1 < t_2 < \ldots < t_k < T$. The number of PMs k is fixed in advance. Each PM carried out at t_i ***reveals and eliminates*** all initial failures existing at the instant t_i. Our goal is to find the optimal location of the PM instants $t_1, \ldots, t_k$ which would minimize the mean number of terminal failures on $[0, T]$.

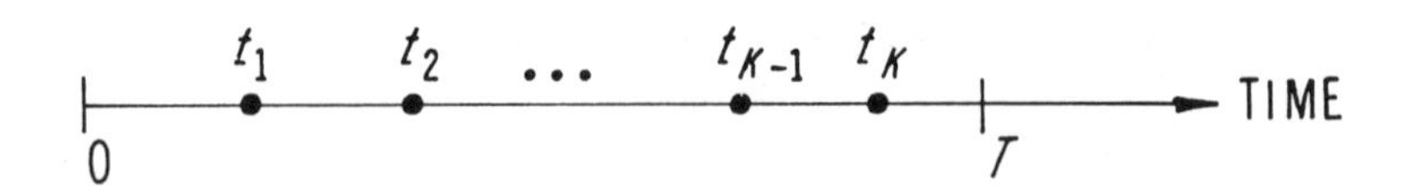

Figure 5.4. Preventive maintenance is scheduled at the instants $t_1, \ldots, t_k$

5.3.3 Some Facts about the Nonhomogeneous Poisson Process

We will need some facts from the theory of NHPP; see Appendix A for a detailed derivation.

Theorem 5.3.1
Let ξ_t be an NHPP process with rate $\lambda(t)$. Denote by $v_k(t_1, t_2)$ the probability that exactly k events appear in the interval $[t_1, t_2]$. Then

$$v_k(t_1, t_2) = \exp[-\Lambda(t_1, t_2)] \frac{[\Lambda(t_1, t_2)]^k}{k!}, \quad k = 0, 1, \ldots, \tag{5.3.1}$$

where $\Lambda(t_1, t_2) = \int_0^{t_2} \lambda(t)dt - \int_0^{t_1} \lambda(t)dt = \int_{t_1}^{t_2} \lambda(t)dt$. In particular, the mean number of events in $[t_1, t_2]$ is

$$E[\xi_{t_2} - \xi_{t_1}] = \Lambda(t_1, t_2), \tag{5.3.2}$$

and the probability of no events in $[t_1, t_2]$ is

$$P(\xi_{t_2} - \xi_{t_1} = 0) = \exp[-\Lambda(t_1, t_2)]. \tag{5.3.3}$$

We see from this theorem that the NHPP is characterized by the integral of the event rate, the so-called cumulative event rate

$$\Lambda(t) = \int_0^t \lambda(v)dv.$$

The key to our analysis is the following theorem (for the proof see Andronov and Gertsbakh 1972; Andronov 1994).

Theorem 5.3.2
The appearance times of terminal failures constitute an NHPP with cumulative event rate $\Lambda_F(t)$ defined as

$$\Lambda_F(t) = \int_0^t \lambda(t-y)F(y)dy. \tag{5.3.4}$$

Often, the following equivalent form is more convenient:

$$\Lambda_F(t) = \int_0^t \lambda(y)F(t-y)dy \ .$$

To derive it, put $y = t - v$ in (5.3.4).

As a corollary, the mean number of terminal failures on $[t_1, t_2]$ is

$$M[t_k, t_{k+1}] = \Lambda_F(t_2) - \Lambda_F(t_1), \tag{5.3.5}$$

and the probability of no terminal failures on $[t_1, t_2]$ is

$$P_0[t_1, t_2] = \exp[-(\Lambda_F(t_2) - \Lambda_F(t_1))]. \tag{5.3.6}$$

5.3.4 Finding the Optimal Location of the PM Points

Suppose that there is only a single PM at $t = t_k$. Then the mean number of terminal failures on $[t_k, t_{k+1} = T]$ is

$$M[t_k, t_{k+1}] = \int_{t_k}^{t_{k+1}} \lambda(v)F(t_{k+1} - v)dv \ . \tag{5.3.7}$$

Indeed, in the absence of the PM, the mean number of failures on $[0, t_{k+1} = T]$ would be $\int_0^{t_{k+1}} \lambda(v)F(t_{k+1} - v)dv$.

The terminal faiulures in $[t_k, t_{k+1}]$ are of two types: those arising from initial failures "born" in $[0, t_k]$ and those arising from initial failures "born" in $[t_k, t_{k+1}]$. The former are *eliminated* by the PM. This is equivalent formally to putting $\lambda(v) = 0$ for $v \in [0, t_k]$. Thus we arrive at (5.3.7).

It is instructive to give a heuristic proof of (5.3.7). Divide the interval $[t_k, t_{k+1}]$ into a large number of small intervals, each of length Δ. An initial failure appears in $[v, v + \Delta]$ with probability $\lambda(v)\Delta$, and with probability $F(t_{k+1} - v)$ it degenerates into a terminal failure *within* the interval $[t_k, t_{k+1}]$. Thus the contribution of one Δ-interval to the mean number of terminal failures is the sum of all terms $\lambda(v)\Delta F(t_{k+1} - v)$, which tends to the integral (5.3.7) when $\Delta \to 0$.

Now suppose that m PMs are planned at the instants $0 < t_1 < t_2 < \ldots < t_m < T$. Then the mean number of terminal failures will be

$$M(t_1, \ldots, t_m) = \sum_{j=0}^{m} \int_{t_j}^{t_{j+1}} \lambda(x)F(t_{j+1} - x)dx, \tag{5.3.8}$$

where $t_0 = 0,\ t_{m+1} = T$.

Now our problem is to locate the PM points $t_1, \ldots, t_m$ optimally in order to minimize $M(t_1, \ldots, t_m)$. This is a typical dynamic programming problem. Let us recall the Bellman's dynamic programming principle (Bellman 1957, p. 18): "An optimal policy has the property that whatever the initial state and the initial decision are, the remaining decisions must constitute an optimal policy with regard to the state resulting from the first decision." We apply this principle to our problem as follows. Denote by $N(x, T;\ s)$ the minimum mean number of terminal failures in the interval $[x, T]$ if

(i) a PM is made at $t = x$;

(ii) s PM points are optimally located in the interval $(x, T]$.

Suppose that we do the first PM in $[x, T]$ at time $t = t_1$, and afterwards follow the optimal strategy in the remaining period $(t_1, T]$. Then we obtain the following principal recurrence relation for $s = 1, 2, \ldots$:

$$N(x, T;\ s + 1) = \min_{x < t_1 < T} \Big(\int_x^{t_1} \lambda(v)F(t_1 - v)dv + N(t_1, T; s) \Big). \tag{5.3.9}$$

It is more convenient to write the integral in (5.3.9) in an equivalent form: $\int_0^{t_1 - x} \lambda(t_1 - y)F(y)dy$ (by means of the change of variables $t_1 - v = y$).

Numerical implementation of Bellman's principle

To facilitate the numerical solution of (5.3.9), we use a discrete time scale. Suppose that a certain interval Δ is taken as a time unit, the interval $[0, T]$ is a multiple of Δ, i.e. $T = \Delta M$, and that the instants for PM t_i can be only the values $t_i = k_i \Delta$, where k_i are integers. Then (5.3.9) takes the form (setting $\Delta = 1$):

$$N(k, M;\ s+1) = \min_{k+1 \le x \le M-(1+s)} \Big(\int_0^{(x-k)} \lambda(x-u) F(u) du + N(x, M;\ s) \Big), \quad (5.3.10)$$

$s = 1, 2, \ldots; k = 0, 1, \ldots, M-2$.

First, we choose the optimum position of a single PM point according to

$$N(k, M;\ 1) = \min_{k+1 \le x \le M-1} \Big(\int_0^{(x-k)} \lambda(x-u) F(u) du + \int_0^{(M-x)} \lambda(M-u) F(u) du \Big). \quad (5.3.11)$$

To find the optimal location of two PM points we use the expressions $N(k, M;\ 1)$ just found:

$$N(k, M;\ 2) = \min_{k+1 \le x \le M-2} \Big(\int_0^{(x-k)} \lambda(x-u) F(u) du + N(x, M;\ 1) \Big), \quad (5.3.12)$$

$k = 0, 1, \ldots, M-3$, and so forth.

Example 5.3.1
The rate of initial failures is $\lambda(t) = 0.3 \times 10^{-2} + 0.75 \times 10^{-3} t$. The time interval has length $T = 50$ time units. On average, there will be one initial failure on $[0, 50]$. The c.d.f. of the regeneration time is $F(t) = 1 - \exp[-0.2t]$, i.e. the initial failure becomes a terminal one after 5 units of time, on average.

The optimal location of a single PM point is at $t = 39$, giving $N(0, 50; 1) = 0.78$. (Recall that this is the mean number of terminal failures.) If we take the PM point in the middle, at $t = 25$, then $N(\cdot) = 1.21$, which is considerably larger.

The optimal location of two PM points is at $t_1 = 32$ and $t_2 = 42$. This gives $N(0, 50; 2) = 0.68$. The optimal location of three PM points is at $t_1 = 27, t_2 = 36, t_3 = 43$. Then $N(0, 50; 3) = 0.6$. The optimal location of four points is at $t_1 = 23, t_2 = 32, t_3 = 39, t_4 = 45$, and the minimum $N(0, 50; 4) = 0.54$. If the arrangement of points were uniform, at $t_1 = 10, t_2 = 20, t_3 = 30, t_4 = 40$, then the mean number of terminal failures would be 0.61.

5.4 Markov-Type Processes with Rewards (Costs)

5.4.1 Markov Chain

The preventive maintenance models considered in this chapter are based on a probabilistic description of the system behavior. This description involves many (more than two) states. An adequate formal tool for finding optimal maintenance policies in this situation is a Markov-type processes with rewards (or costs) associated with the transitions from state to state.

Our starting point will be a finite-state Markov chain. Although we asume that the reader is already familiar with the notion of a Markov chain, let us recall briefly the relevant facts.

Consider a sequence of random variables $\{\xi_k,\ k = 0, 1, 2, \ldots\}$, where each ξ_k takes on a finite number of possible values which we call ***states***. These possible values will be denoted by the set of nonnegative integers from the set $S = \{1, 2, \ldots, n\}$. If $\xi_k = i$, then the Markov chain is said to be in state i at time k. We will use often the word "process" instead of "Markov chain."

The random mechanism which governs the transitions from state to state is the following. Suppose that whenever the process is in state i, there is a probability P_{ij} that the next state will be j. Formally, we assume that

$$P(\xi_{n+1} = j|\xi_n = i,\ \xi_{n-1} = i_{n-1}, \ldots, \xi_1 = i_1, \xi_0 = i_0) = P_{ij}. \tag{5.4.1}$$

P_{ij} is called the one-step transition probability. This is the probability that the process makes a transition from state i to state j. Expression (5.4.1) postulates that this probability *does not* depend on the states of the chain before it entered state i. Obviously, $P_{ij} \geq 0$ and $\sum_{j=1}^{n} P_{ij} = 1$. We will need the $n \times n$ matrix $\mathbf{P}$ of one-step transition probabilities:

$$\mathbf{P} = ||P_{ij}||,\ \ i, j = 1, 2, \ldots, n\ . \tag{5.4.2}$$

One can imagine the evolution in time of a Markov chain ξ_k as the movement of a fictitious particle over a set of states. This movement is a sequence of jumps, every jump occurs at time instants $k,\ k = 1, 2, 3, \ldots$. The particle is in state i at time n. At time $n + 1$, it jumps from i to j with probability P_{ij}, *no matter where the particle was before time n*. This is in fact the Markovian property expressed by (5.4.1).

We now define the r-step transition probabilities. Let P_{ij}^r be the probability that the process which presently is in state i will be in state j after r transitions:

$$P_{ij}^r = P(\xi_{q+r} = j|\xi_q = i),\ r > 0,\ i, j = 1, 2, \ldots, n\ . \tag{5.4.3}$$

Denote by $\mathbf{P}^{(r)}$ the matrix of the r-step transition probabilities. Then

$$\mathbf{P}^{(r)} = ||P_{ij}^r|| = \mathbf{P}^r. \tag{5.4.4}$$

Thus the r-step transition probabilities are obtained simply by raising the **P** matrix to the rth degree. This result is known as the Chapman–Kolmogorov equation; see Taylor and Karlin (1984).

We will consider a ***nonperiodic*** Markov chain with a single class of communicating states. Formally, this is equivalent to the assumption that for any pair of states i and j, there is some s such that $P_{ij}^r > 0$ for all $r \geq s$. In other words, there is a positive probability that a particle can reach any state j from any state i in s transitions.

The following principal theorem establishes the limiting behavior of the r-step transition probabilities.

Theorem 5.4.1
(i) $\lim_{r\to\infty} P_{ij}^{(r)} = \pi_j$; π_j are nonnegative, and $\sum_{j=1}^n \pi_j = 1$.

(ii) π_j is the unique nonnegative solution of the system

$$\pi_j = \sum_{i=1}^n \pi_i P_{ij},\ j = 1, 2, \ldots, n,\ \sum_{i=1}^n \pi_i = 1. \tag{5.4.5}$$

The π_i are termed ***stationary*** or ***limiting*** probabilities of the Markov chain.

Let us mention several valuable properties of the stationary probabilities. By Theorem 5.4.1, π_i is the limiting probability that the process is in state i at some remote time instant t, formally as $t \to \infty$ (assume that each transition takes one time unit). Suppose that we observe the chain during a large number of transitions N. Let $K_i(N)$ be the number of times the process visits state i. Then the long-run proportion $K_i(N)/N \to \pi_i$ as $N \to \infty$.

For state j, define m_{jj} to be the mean number of transitions until a Markov chain, starting in state j, returns again to that state. It turns out that

$$\pi_j = \frac{1}{m_{jj}}. \tag{5.4.6}$$

A heuristic proof is as follows (see e.g. Ross 1993). The chain spends 1 unit of time in state j each time it visits j. Since, on average, it visits j once in m_{jj} steps, the limiting probability for this state must be $1/m_{jj}$.

The returns of the Markov chain to any fixed state, say state 1, form a renewal process. The behavior of the chain after its return to state 1, which took place at the time instant K_1, does not depend on the past history for $t < K_1$. If the returns to state 1 take place at the instants $K_1, K_2, K_3, \ldots$, then the intervals $K_2 - K_1, K_3 - K_2, \ldots$, etc. are i.i.d. r.v.s with mean m_{11}.

5.4.2 Semi-Markov Process

It was assumed that the transitions in the Markov chain take place at the instants $t = 1, 2, 3, \ldots$. One might imagine that after the transition into state

i, the Markov chain is "sitting" in this state i exactly one unit of time, after which the next transition takes place.

Let us consider now a very useful generalization of a Markov chain called semi-Markov process (SMP). In simple terms, this is a chain with *random* time intervals between transitions; these random intervals do not depend on the past trajectory.

Suppose that we have the same set of states $\{1, 2, \ldots, n\}$ as for the Markov chain and the same probabilistic mechanism governing the change of states. Suppose that our new process starts at $t = 0$ from state i. The next state will be, as in the Markov chain, the state j chosen with probability P_{ij}. Now introduce the one-step ***transition time*** τ_{ij}. If the next state is j, then the SMP spends time τ_{ij} in i, and then jumps into j. The SMP will be denoted by ζ_t. The distribution of τ_{ij} is defined as

$$P(\tau_{ij} \le t | \text{transition } i \Rightarrow j) = F_{ij}(t). \tag{5.4.7}$$

The trajectory of ζ_t is constructed as follows. Take the initial state i. Choose randomly the next state according to the distribution P_{ik}, $k = 1, 2, \ldots, n$. Suppose the next state is s. Then let the process stay at i a random time $\tau_{is} \sim F_{is}(t)$, after which the process jumps into state s. If the next state is q the process sits in s a random time $\tau_s \sim F_{sq}(t)$ and then jumps into q, etc. Figure 5.5 shows the trajectory of an SMP.

Denote by Q_j the limiting probability of state j. Formally,

$$Q_j = \lim_{t \to \infty} P(\zeta_t = j) . \tag{5.4.8}$$

Q_j equals the long-run proportion of time which the SMP spends in state j:

$$Q_j = \lim_{t \to \infty} \frac{\text{time in } j \text{ on } [0, t]}{t} . \tag{5.4.9}$$

Denote by L_{jj} the mean first-passage (return) time from state j to state j.

To formulate the next theorem, we need to introduce the mean one-step transition time ν_i for state i. From the description of ζ_t it follows that

$$\nu_i = \sum_{j=1}^{n} P_{ij} \int_0^\infty x dF_{ij}(x). \tag{5.4.10}$$

The following important theorem expresses Q_j and L_{jj} as a function of ν_j and π_j; see Ross (1970).

Theorem 5.4.2

$$Q_j = \frac{\pi_j \nu_j}{\sum_{i=1}^{n} \pi_i \nu_i}, \tag{5.4.11}$$

$$L_{jj} = \frac{\sum_{i=1}^{n} \pi_i \nu_i}{\pi_j} . \tag{5.4.12}$$

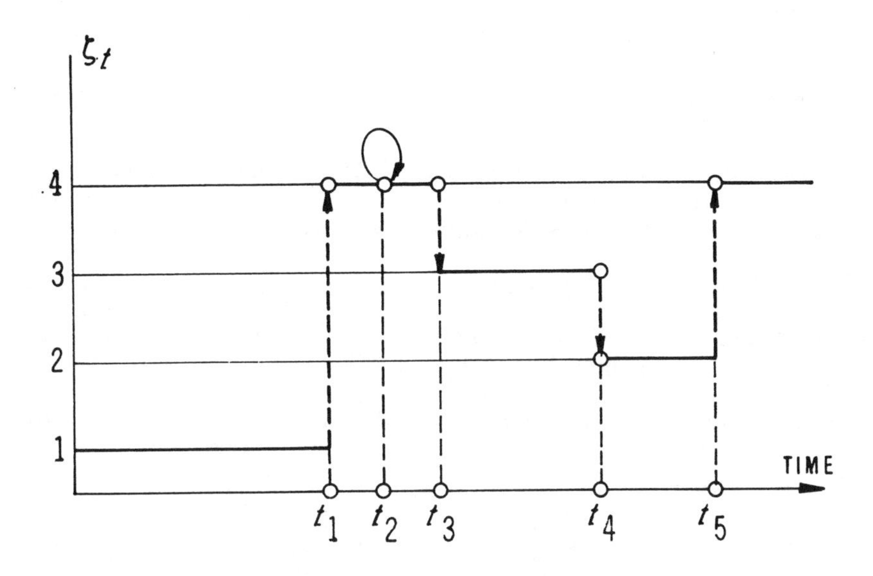

Figure 5.5. Trajectory of an SMP. Transitions occur at t_1, t_2, t_3, t_4, t_5 into the states $4, 4, 3, 2$ and 4, respectively.

Let us present a heuristic proof of this theorem. Consider a large time span $[0, T]$, during which K_i transitions occur into state i. The SMP spends, on average, time ν_i in state i after each such transition. Therefore, $T \approx \sum_{i=1}^{n} K_i \nu_i$. Since the total time spent in j is $K_j \nu_j$, $Q_j \approx K_j \nu_j / \sum_{i=1}^{n} K_i \nu_i$. Now divide both sides by $K = \sum_{i=1}^{n} K_i$ and take into account that $K_i / K \approx \pi_i$. This explains the first claim. Suppose that ζ_t returns into j every L_{jj} units of time. On average, after each return the process sits in j time ν_j. Thus, $Q_j \approx \nu_j / L_{jj}$, which explains the second claim of the theorem.

Further details on the SMP can be found in Ross (1970) or Gertsbakh (1977).

5.4.3 SMP with Rewards (Costs)

Now let us define *rewards* or *costs* associated with the transitions in the SMP. Assume that a reward $\psi_{ij}(\tau_{ij})$ is accumulated for a transition $i \Rightarrow j$ which lasts τ_{ij}. Then the average reward ω_i for a one-step transition from i is obtained by averaging $\psi_{ij}(x)$ over all possible durations of the one-step transition and over all possible destinations j:

$$\omega_i = \sum_{j=1}^{n} P_{ij} \int_0^{\infty} \psi_{ij}(x) dF_{ij}(x) . \tag{5.4.13}$$

Now we are ready for a formula for long-run mean reward per unit time.

Theorem 5.4.3
The mean long-run reward per unit time is

$$g = \frac{\sum_{i=1}^{n} \pi_i \omega_i}{\sum_{i=1}^{n} \pi_i \nu_i}. \tag{5.4.14}$$

It is instructive to derive this formula heuristically. Consider again a large time span $[0, T]$. During this time, K_i transitions occur into i, and after each such transition the SMP stays in this state, on average, for time ν_i. Therefore,

$$T \approx K_1\nu_1 + K_2\nu_2 + \ldots + K_n\nu_n. \tag{5.4.15}$$

On the other hand, every visit to state i gives a reward whose mean value is ω_i. Therefore, the total accumulated reward on $[0, T]$ is

$$\text{Reward on } [0, T] \approx K_1\omega_1 + K_2\omega_2 + \ldots + K_n\omega_n. \tag{5.4.16}$$

Now

$$g \approx \frac{\text{Reward on } [0, T]}{T} = \Big(\sum_{i=1}^{n} \frac{K_i\omega_i}{\sum_{j=1}^{n} K_j}\Big)\Big(\sum_{i=1}^{n} \frac{K_i\nu_i}{\sum_{j=1}^{n} K_j}\Big)^{-1}. \tag{5.4.17}$$

Now note that $K_i / \sum_{i=1}^{n} K_i$ is approximately equal to the stationary probability π_i. This formula will be used in the next two sections.

5.4.4 Continuous-Time Markov Process

Let $\xi(t)$ be a continuous-time Markov process with a finite number of states $0, 1, 2, \ldots, N$. This means that for all t_1, t_2, $t_1 \le t_2$, and any pair of states i, j,

$$\begin{aligned} &P(\xi(t_1 + t_2) = j | \xi(t_1) = i, \text{any history of } \xi(u) \text{ on } [0, u], u < t_1) \\ &= P(\xi(t_1 + t_2) = j | \xi(t_1) = i) . \end{aligned} \tag{5.4.18}$$

It is postulated, therefore, that the future behavior of $\xi(t)$ after t_1 depends only on its present state i at t_1.

A convenient way of defining the transition probabilities $P_{ij}(t_2) = P(\xi(t_1 + t_2) = j | \xi(t_1) = i)$ is to describe the behavior of the process on a small time interval.

Let us define the following *transition rates* λ_{kj}: as $\Delta t \to 0$,

$$\begin{aligned} &P(\xi(t + \Delta t) = j | \xi(t) = k) = \lambda_{kj}\Delta t + \mathrm{o}(\Delta t),\ j \ne k, \\ &P(\xi(t + \Delta t) = j | \xi(t) = j) = 1 - \sum_{j \ne k} \lambda_{kj}\Delta t + \mathrm{o}(\Delta t) . \end{aligned} \tag{5.4.19}$$

Let the initial state at $t = 0$ be i: $P(\xi(0) = i) = 1$. We will describe a standard procedure for finding the transition probabilities

$$P_{ij}(t) = P(\xi(t) = j | \xi(0) = i) .$$

Let us introduce the following matrix $\mathbf{B}$:

$$\mathbf{B} = \begin{bmatrix} \lambda_0^* & \lambda_{01} & \lambda_{02} & . & . & . & \lambda_{0N} \\ \lambda_{10} & \lambda_1^* & \lambda_{12} & . & . & . & \lambda_{1N} \\ . & . & . & . & . & . & . \\ . & . & . & . & . & . & . \\ . & . & . & . & . & . & . \\ \lambda_{N0} & \lambda_{N1} & \lambda_{N2} & . & . & . & \lambda_N^* \end{bmatrix}.$$

Here

$$\lambda_k^* = -\sum_{j \neq k} \lambda_{kj}, \quad k = 0, 1, \ldots, N. \tag{5.4.20}$$

Let $\mathbf{P}_i'(t)$ and $\mathbf{P}_i(t)$ be the columns with elements $P_{i0}'(t), \ldots, P_{iN}'(t)$ and $P_{i0}(t), \ldots, P_{iN}(t)$, respectively. The transition probabilities $P_{ij}(t)$ satisfy the following system of differential equations:

$$\mathbf{P}_i'(t) = \mathbf{B}'\mathbf{P}_i(t), \tag{5.4.21}$$

with the initial condition $P_{ik}(0) = 0$, if $k \neq i$ and $P_{ii}(0) = 1$. $\mathbf{B}'$ is the transpose of $\mathbf{B}$.

To prove (5.4.21), let us derive the equation for $P_{ij}'(t)$. Expression (5.4.21) says that

$$P_{ij}'(t) = \sum_{k \neq j} P_{ik}(t)\lambda_{kj} + \lambda_j^* P_{ij}(t). \tag{5.4.22}$$

Let us consider the transition probability from state i to state j during the interval $[t, t + \Delta t]$. To be in state j at $t + \Delta t$, the process $\xi(t)$ either has to be in state k, $k \neq j$, at time t and then jump into state j during the interval $[t, t + \Delta t]$, or has to be already in state j at t and remain in this state during the interval $[t, t + \Delta t]$. This leads to the following probability balance equation:

$$P_{ij}(t + \Delta t) = \sum_{k \neq j} P_{ik}(t)\lambda_{kj}\Delta t + \big(1 - \sum_{j \neq k} \lambda_{kj}\Delta t\big)P_{ij}(t) + o(\Delta t). \tag{5.4.23}$$

Now transfer the term $P_{ij}(t)$ from the right to the left, divide by Δt and set $\Delta t \to 0$. We obtain the desired equation (5.4.22).

The method which usually simplifies the analytic solution of the system (5.4.21) is applying the Laplace transform, which leads to a system of linear algebraic equations. Let

$$\pi_j(s) = \int_0^\infty e^{-st} P_{ij}(t)dt. \tag{5.4.24}$$

Applying the Laplace transform (5.4.24) to each row of the system (5.4.21) and using the formula

$$\int_0^\infty e^{-st} P_{ij}'(t)dt = s\pi_j(s) - P_{ij}(0), \tag{5.4.25}$$

we obtain the system

$$s\Pi(s) - \mathbf{P}_i(0) = \mathbf{B}'\Pi(s), \tag{5.4.26}$$

where $\Pi(s)$ is the column consisting of elements $\pi_i(s)$, $j = 0, 1, \ldots, N$. From this follows the system of equations

$$(\mathbf{I}s - \mathbf{B}')\Pi(s) = \mathbf{P}_i(0), \tag{5.4.27}$$

where $\mathbf{I}$ is the identity matrix. This system must be solved with respect to $\pi_j(s)$, $j = 0, \ldots, N$, and the functions $P_{ij}(t)$ must be found with the help of the Laplace transform tables; see Appendix C.

For a better understanding of the process $\xi(t)$ defined via the transition rates, let us describe its sample paths. Suppose that $\xi(0) = i_1$. Then $\xi(t)$ stays in i_1 a random time $\tau_{i_1} \sim \text{Exp}(-\lambda^*_{i_1})$, after which it jumps into state i_2 which is chosen with probability $\lambda_{i_1,i_2}/(-\lambda^*_{i_1})$. In state i_2, the process stays a random time $\tau_{i_2} \sim \text{Exp}(-\lambda^*_{i_2})$, after which it jumps into a new state i_3, etc.

5.5 Opportunistic Replacement of a Two-Component System

5.5.1 General Description

A system consists of two independent parts (components), named a and b. Both parts operate continuously and fail independently. If a component fails, an inspection will reveal it and the failed component will be replaced. After replacement, it is as good as new. If one component, a or b, has failed and is replaced, then there are two options with regard to the other component: to replace it too or not. The first option is what we call "opportunistic replacement."

In Sect. 4.1 we considered an example of an opportunistic replacement scheme in which both components were *always* replaced upon the failure of either. This policy turned out to be less costly than the policy of replacing each component separately (Example 4.1.3). A natural extension of this simplified opportunistic replacement is the following: when one of the two components has failed, replace the second one only if its age exceeds some critical value $T_{\max}$.

We shall assume that the system is periodically inspected, and that the inspection always reveals any existing failure. The following four situations may occur:
(1) no failure discovered, no action taken;
(2) one part has failed, the other has not, and only the failed part is replaced;
(3) one part has failed, the other has not, and both parts are replaced;
(4) both parts found failed, and both are replaced.
Let us introduce the costs associated with these situations. There is a zero cost associated with (1); the cost of (2), the single repair cost, is C_f; the cost of (3) or (4), the joint repair cost, is $C_{p,f}$.

5.5.2 Formal Setting of the Problem

The time interval between inspections Δ is set to 1. All inspections and repairs take negligible time. The lifetimes of a and b are τ_a and τ_b, respectively, and they are independent discrete random variables on a unit-step grid. Let

$$P(\tau_a = i) = p_i,\ i = 1, \dots, K;\ P(\tau_b = j) = q_j,\ j = 1, \dots, K. \tag{5.5.1}$$

We will need the conditional probabilities that a part fails at "age" k, $k > r$, if it has survived age r. Let

$$P(\tau_a = i | \tau_a > r) = p_i^{(r)},\quad i = r+1, r+2, \dots;$$
$$P(\tau_b = i | \tau_b > r) = q_i^{(r)},\quad i = r+1, r+2, \dots$$

Obviously,

$$p_i^{(r)} = \frac{p_i}{1 - \sum_{j=1}^r p_j};\ q_i^{(r)} = \frac{q_i}{1 - \sum_{j=1}^r q_j}. \tag{5.5.2}$$

Our opportunistic replacement policy is formulated as follows: if during an inspection it has been revealed that one of the two parts has failed, then *replace the second part only if its age exceeds the critical value* $T_{\max} = N$.

The state ζ of the two-part system will be defined as follows. Put $\zeta = (K_a, K_b)$, where K_a and K_b are the respective component ages after the inspection *and replacement*. The consecutive states of the system will be denoted by $\zeta_0, \zeta_1, \dots$.

Consider an example. Time t is measured in integers $0, 1, 2, \dots$ Suppose that $N = 2$. Both parts are brand new at $t = 0$; the lifetimes of a and b are $\tau_a = 1, \tau_b = 4$. The system starts at state $\zeta_0 = (0, 0)$ at $t = 0$ (think of this as signifying that at $t = 0$ both parts have been replaced). The first failure of a takes place at $t = 1$, and only a is replaced since the age of b is less than $N = 2$. So, after the first replacement, the state of the system is defined as $\zeta_1 = (0, 1)$. Suppose that the part which replaces the failed a-part has lifetime $\tau_a^* = 4$. The next failure takes place at $t = 4$; this is the failure of b. The age of a is now $3 > N = 2$. Thus, both components need to be replaced, and the state of the system becomes $\zeta_2 = (0, 0)$. Thus the trajectory of ζ_j is $(0, 0) \Rightarrow (0, 1) \Rightarrow (0, 0) \Rightarrow \dots$.

Suppose that $\tau_a = \tau_b = 4$. Then the initial state is (0,0). Failure takes place for the first time at $t = 4$, both components are replaced, and the next state is again (0,0). So, the trajectory is $\zeta_0 = (0, 0) \Rightarrow \zeta_1 = (0, 0) \Rightarrow \dots$.

ζ_j has the following state space: $S = \{(0, 0); (0, j),\ j = 1, \dots, N; (i, 0),\ i = 1, \dots N, \}$. The state (i, i) for $i > 0$ is impossible. Indeed, after replacement, either or both ages will be zero.

It is an important fact that the states of the system ζ_j, $j = 0, 1, 2, \dots$, form a Markov chain. Indeed, the probability of the transition $(i_1, j_1) \Rightarrow (i_2, j_2)$ depends only on the present state (i_1, j_1) and does not depend on previous states. This is due to the definition of the system state. If the ages of a and b are fixed, the future behavior is determined via the conditional probabilities (5.5.2).

Our goal is to obtain an expression for the average long-run costs per unit time. For this purpose we use the technique described in the previous section. In fact, the intervals between transitions from state to state are random, and ζ_t fits the description of a semi-Markov process (Section 5.4.2).

We proceed with a numerical example.

5.5.3 Numerical Example

The following are component lifetime data:

i	p_i	$p_i^{(1)}$	$p_i^{(2)}$	q_i	$q_i^{(1)}$	$q_i^{(2)}$
1	0.08	–	–	0.05	–	–
2	0.14	0.152	–	0.08	0.084	–
3	0.17	0.185	0.218	0.16	0.168	0.184
4	0.17	0.180	0.218	0.30	0.316	0.345
5	0.15	0.160	0.192	0.20	0.211	0.230
6	0.11	0.120	0.141	0.10	0.105	0.115
7	0.18	0.203	0.231	0.11	0.116	0.126

It will be assumed that $N = 2$, and thus the system states are : (0,0), (0,1), (0,2), (1,0), (2,0).

The most tedious part of this example is calculating the transition probabilities. Let us start with state (0,0). To avoid mistakes, it is advisable to picture the sample space of (τ_a, τ_b); see Fig. 5.6.

A point with coordinates (r, s) represents the event $(\tau_a = r, \tau_b = s)$ and since the lifetimes of a and b are independent, its probability is $p_r q_s$.

If the event $A_{(1,0)}$ takes place, the next state after (0,0) will be (1,0). (Verify it!) The event $A_{(0,1)}$ denotes transition into (0,1). Similar meanings attach to the events $A_{(0,2)}$, $A_{(2,0)}$. The remaining probability mass corresponds to the transition $(0,0) \Rightarrow (0,0)$. It is a matter of a routine computation to obtain $p_{00,01} = 0.046$; $p_{00,02} = 0.122$; $p_{00,10} = 0.076$; $p_{00,20} = 0.062$; $p_{00,01} = 0.694$.

Denote by ν_{00} the mean transition time from state (0,0). If the event $B_1 = (\tau_a = \tau_b = 1) \bigcup A_{(1,0)} \bigcup A_{(0,1)}$ takes place, then the system stays in (0,0) one time unit. If $B_2 = (\tau_a = \tau_b = 2) \bigcup A_{(2,0)} \bigcup A_{(0,2)}$ takes place, the transition lasts two time units. The events A_3, A_4, A_5, A_6 and A_7 mean that the transition times are $3, 4, 5, 6$ and 7, respectively; see Fig. 5.6. So, $\nu_{00} = P(B_1) + 2P(B_2) + 3P(A_3) + \ldots + 7P(A_7)$. The result is $\nu_{00} = 3.247$.

Let us show how to compute the transition characteristics from state (2,0). Figure 5.7 contains the necessary information. The τ_a axis is now the ***residual*** life of a given that it survived past $t = 2$; see the column $p_i^{(2)}$ in the table above. The events $A_{(0,1)}$ and $A_{(0,2)}$ correspond to the transition into (0,1) and (0,2), respectively.

Indeed, suppose that $\zeta_i = (2,0)$, i.e. b is brand new and a has age 2. $\zeta_{i+1} = (0,1)$ means that after the failure of a, b has age 1. The transition $\zeta_i \Rightarrow \zeta_{i+1}$ can happen if and only if the residual lifetime of a is 1 and the age of b is greater than 1, exactly as defined by the event $A_{0,1}$. This gives $p_{20,01} = 0.207$; $p_{20,02} = 0.190$; $p_{20,00} = 0.603$.

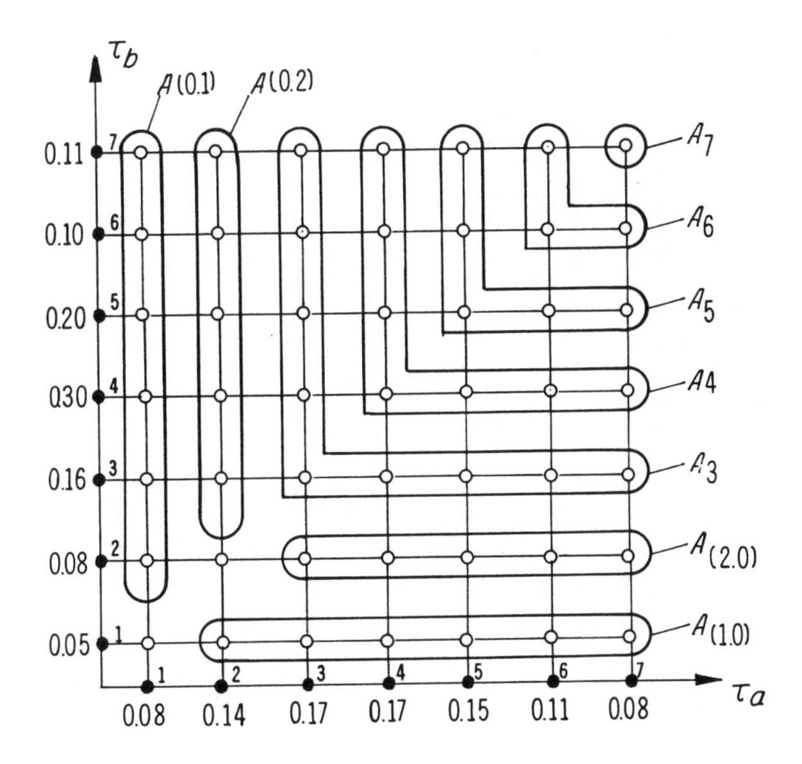

Figure 5.6. The sample space for the transitions from (0,0)

The events B_1 through B_5 (see Fig. 5.7), correspond to the transition times $1, 2, \ldots, 5$ respectively. Thus we arrive at $\nu_{20} = 2.592$.

We omit similar calculations for the transitions starting in states (0,1), (1,0) and (0,2). We eventually obtain the following transition probability matrix **P**:

state	$(0,0)$	$(0,1)$	$(1,0)$	$(2,0)$	$(0,2)$
$(0,0)$	0.694	0.046	0.076	0.062	0.122
$(0,1)$	0.719	0	0.077	0.131	0.073
$(1,0)$	0.653	0.144	0	0.042	0.161
$(2,0)$	0.603	0.207	0	0	0.190
$(0,2)$	0.562	0	0.169	0.269	0

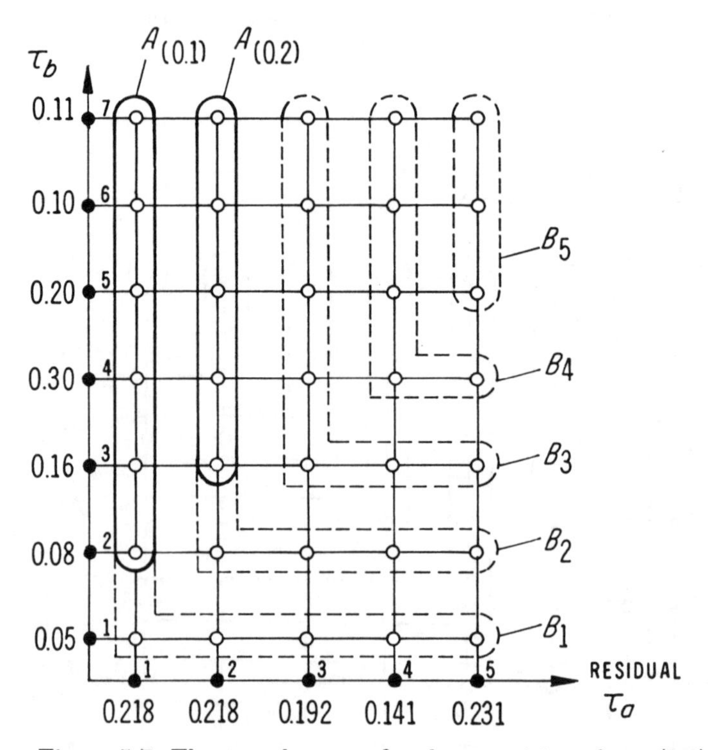

Figure 5.7. The sample space for the transitions from (2,0)

By solving (5.4.5) (see Theorem 5.4.1) we find the stationary probabilities $\pi_{00} = 0.670, \pi_{01} = 0.059, \pi_{10} = 0.075, \pi_{20} = 0.083, \pi_{02} = 0.113$.

The last task is to find the mean one-step costs. Let us assume that a single component replacement costs $C_f = \$1000$, and two-component replacement costs $C_{p,f} = \$1400$. Suppose that the system state is (0,0). The transition into state (0,0) costs $C_{p,f}$. The transitions to any other state cost C_f. Therefore, the mean cost associated with one transition starting (0,0) is

$$\omega_{00} = P\big((0,0) \Rightarrow (0,0)\big)C_f + \big(1 - P((0,0) \Rightarrow (0,0))\big)C_{p,f}. \tag{5.5.3}$$

The mean cost of transitions starting from other states is computed similarly. For example,

$$\omega_{01} = P\big((0,1) \Rightarrow (0,0)\big)C_f + \big(1 - P((0,1) \Rightarrow (0,0))\big)C_{p,f}. \tag{5.5.4}$$

The following table summarizes the stationary probabilities, mean transition times and costs.

state (i,j)	$(0,0)$	$(0,1)$	$(1,0)$	$(2,0)$	$(0,2)$
π_{ij}	0.670	0.059	0.075	0.083	0.113
ν_{ij}	3.247	3.112	2.870	2.592	2.321
ω_{ij}	1278	1288	1261	1241	1225

Now everything is ready to compute the mean costs in \$ per unit time. By (5.4.14),

$$g = \frac{\sum \pi_{ij}\omega_{ij}}{\sum \pi_{ij}\nu_{ij}} = 375.6. \tag{5.5.5}$$

By way of comparison, let us calculate the costs for an alternative policy: replace both units at the failure of either. The result is

$$g_1 = C_{p,f}/E[\min(\tau_a, \tau_b)] = 431.$$

5.6 Discovering a Malfunction in an Automatic Machine

5.6.1 Problem Description

An automatic machine produces one article every time unit. If the machine is in a "good" state, then there is the probability p_0 that the article produced is defective. If the machine has a malfunction (i.e. it is in a failure state) this probability is p_1 and $p_1 > p_0$. The entire machine operation time is divided into periods during which N articles are produced. It is assumed that failure can occur with constant probability γ at the beginning of each period. The state of the machine is tested by taking samples of n articles at the end of each period from the batch of N articles produced during this period. All articles in a sample are examined and if the number X of defective articles is k or more, there is deemed to be a malfunction (k is the prescribed "critical" level). If it turns out that $X \geq k$, the machine is stopped for a time m_p, during which it is established whether there is a failure or not. If the alarm was false, the machine immediately resumes operation. If not, an additional time m_a is spent carrying out machine repair, after which the machine starts operation with the initial low level p_0 of defective rate.

It is assumed that each good article gives a reward of C_g and each defective article incurs a penalty of C_d. Our problem is to find the optimal "stopping" rule $X \geq k$, $k = 1, 2, \ldots, n$, which would *maximize* the reward per unit time.

5.6.2 Problem Formalization

Define an SMP with the following states: the machine is operating, there is no malfunction (state 1); the machine is operating (i.e. producing articles), there is a malfunction (state 2); the machine is down, it is repaired after a malfunction has been discovered (state 3); the machine is not operating, but there is no malfunction (state 4).

From the above description it follows that the transition probabilities $P_{31} =$ probabilities $P_{31} = P_{41} = P_{23} = 1$; see Fig. 5.8.

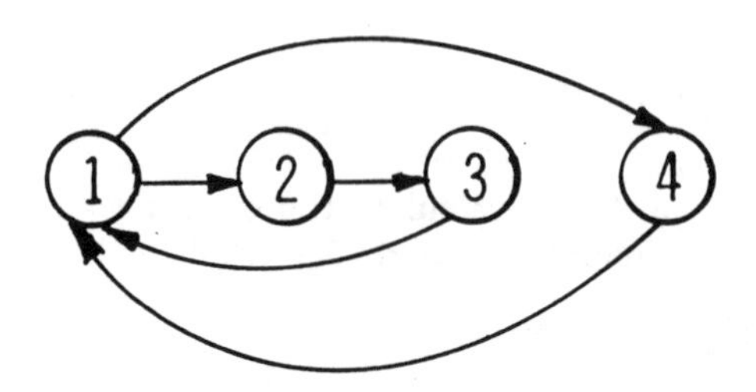

Figure 5.8. State transition diagram

Let $b_0(n,k)$ be the probability that of a sample of n articles k or more are defective if the machine is in order, and let $b_1(n,k)$ be the same probability for a machine which has a malfunction.

The number of defective articles in a random sample of size n taken from a batch of size N has a so-called hypergeometric distribution; see e.g. Devore (1982, p 109). For $N \gg n$ and for p_0 and p_1 not too close to zero, which is what we have, this distribution can be very well approximated by a binomial $X \sim B(n,p)$. Then we can write

$$b_0(n,k) = P(X \geq k; p_0) = \sum_{i=k}^{n} \binom{n}{i} p_0^i (1-p_0)^{n-i}, \tag{5.6.1}$$

$$b_1(n,k) = P(X \geq k; p_1) = \sum_{i=k}^{n} \binom{n}{i} p_1^i (1-p_1)^{n-i}. \tag{5.6.2}$$

Let Y be the ordinal number of the period at the beginning of which the malfunction occurs. According to the problem description, $Y \sim Geom(\gamma)$:

$$P(Y=q) = (1-\gamma)^{q-1}\gamma, \quad q \geq 1.$$

It follows that

$$P(Y > m) = \sum_{q=m+1}^{\infty} P(Y=q) = (1-\gamma)^m. \tag{5.6.3}$$

The transition $1 \Rightarrow 4$ occurs if and only if the alarm signal was received before the malfunction appeared. By (5.6.1) and (5.6.3) it happens at the end of the mth period with probability

$$p_{14}(m) = [1 - b_0(k,n)]^{m-1} b_0(k,n)(1-\gamma)^m. \tag{5.6.4}$$

From this it follows that

$$P_{14} = \sum_{m=1}^{\infty} p_{14}(m) = b_0(k,n)(1-\gamma)[1-(1-b_0(k,n))(1-\gamma)]^{-1}. \tag{5.6.5}$$

Obviously, $P_{12} = 1 - P_{14}$. Now we have the following transition probability matrix $\mathbf{P}$:

state	1	2	3	4
1	0	P_{12}	0	P_{14}
2	0	0	1	0
3	1	0	0	0
4	1	0	0	0

The system (5.4.5), $(\pi_1, \dots, \pi_4) = (\pi_1, \dots, \pi_4)\mathbf{P}$, has the following solution:
$\pi_1 = D$, $\pi_2 = D \cdot P_{12}$, $\pi_3 = D \cdot P_{12}$, $\pi_4 = D \cdot P_{14}$, where $D = 1/(3 - P_{14})$.

We now calculate the mean one-step transition times. Obviously, $\nu_3 = m_p + m_a$, $\nu_4 = m_p$. The machine stays in state 1 a random number of periods Z, and this number exceeds m if and only if there was no malfunction and no false alarm during these m periods. Thus,

$$P(Z > m) = (1 - b_0(k,n))^m (1 - \gamma)^m, \quad m > 1. \tag{5.6.6}$$

It is a matter of routine calculation to show that $E[Z] = [(1 - b_0(k,n))(1-\gamma)]^{-1}$, and therefore the mean one-step transition time from state 1 is $\nu_1 = NE[Z]$. If the machine is in state 2, the probability of discovering the malfunction after one period is $b_1(k,n)$. It is easy to derive from here that the machine stays in state 2, on average, for time $\nu_2 = N/b_1(k,n)$.

It remains to determine the mean one-step rewards. Following the problem description, we obtain

$\omega_1 = \nu_1[(1 - p_0)C_g - p_0 C_d]$,
$\omega_2 = \nu_2[(1 - p_1)C_g - p_1 C_d]$,
$\omega_3 = \omega_4 = 0$.

Indeed, an operating machine produces in state 1 a good unit with probability $1 - p_0$, and a defective one with probability p_0. This explains the formula for ω_1. The formula for ω_2 is similar.

Now we have all the ingredients to allow us to use formula (5.4.14) for the mean long-run reward per unit time:

$$g(k) = \frac{\sum_{i=1}^4 \pi_i \omega_i}{\sum_{i=1}^4 \pi_i \nu_i} . \tag{5.6.7}$$

5.6.3 Numerical Example

Let us investigate expression (5.6.7) for the following values of the parameters involved: $\gamma = 0.1$; $N = 100$ (on average, a malfunction appears once per 1 000 articles); $m_p = 50, m_a = 100$.

The probabilities of producing a defective article are $p_0 = 0.05$ and $p_1 = 0.2$. This means that, on average, one out of 20 articles is defective when the machine operates normally, and one out of 5 is defective when there is a malfunction.

The sample size for checking the machine state is $n = 20$. The reward $C_g = 1$. The penalty for a defective article is $C_d = 1$.

Let us investigate the expression for $g(k)$ numerically. Figure 5.9 presents the results. The optimal $k = 5$. This guarantees $g(5) = 0.76$.

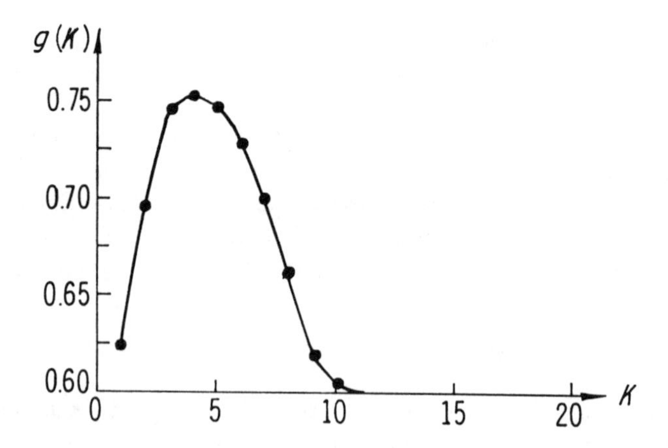

Figure 5.9. Reward as a function of the critical number k

Smaller or larger values of k reduce the reward substantially, as Fig. 5.9 shows. For example, if we take $k = 2$, there will be too many false alarms; if $k = 10$ then the machine will produce too many defective articles.

5.7 Preventive Maintenance of Objects with Multidimensional State Description

5.7.1 General Description

We have considered so far in this chapter several preventive maintenance models based on the observation of the state parameter. One such parameter was the number of failed lines in the multiline system. Another was the state ζ_t of the two-component system described in Sect. 5.5. The system degradation in the two-stage failure model (Sect. 5.3) is reflected by the accumulated number of initial failures.

The main difficulty intrinsic to these and similar problems is finding an appropriate stochastic description of the process which reflects the system degradation (deterioration) or the damage accumulation. The situation is relatively simple if this description can be given in the terms of a one-dimensional Markov-type process.

In practice, however, a situation where the state of the system (component) is described well enough by a *one-dimensional* parameter is the exception rather than the rule. More often an adequate description of the state of a technical object is given in terms of some *multidimensional* process

$$\eta(t) = (\eta_1(t), \eta_2(t), \ldots, \eta_n(t)). \tag{5.7.1}$$

An important practical issue is how to construct, based on the observations of the vector process $\eta(t)$, an optimal preventive maintenance procedure. There are several cases where in fact we can reduce the situation to one dimension.
(a) The processes $\eta_k(t)$, $k = 2, \ldots, n$, are functionally related to the first coordinate $\eta_1(t)$ as $\eta_k(t) = \psi(\eta_k(t)) + \epsilon_t$, where ϵ_t is an observation noise. In this case actually only $\eta_1(t)$ contains useful information.
(b) All $\eta_i(t)$ are independent processes which can be observed, monitored and controlled independently. For example, in a car the state of the braking system is described by the average wear on the brake disk $\eta_1(t)$, and the state of the engine by the fuel consumption index $\eta_2(t)$. We might assume that there is no connection between these two processes. One can design maintenance procedures separately for the braking system and for the engine. The only connection between the maintenance actions for both these systems may arise as a result of cost reduction when these two subsystems undergo preventive maintenance simultaneously.

However, in general, the above-described simplifications are not possible, and we have the rather difficult problem of controlling a multi-dimensional process. The complications arise as a result of dealing with a multi-dimensional "critical region", which replaces the one-dimensional "critical" level. Besides, and this is even more important, we need an adequate probabilistic description of a multidimensional stochastic process. The difficulties in arriving at this description are further aggravated by the fact that we never observe "pure", undisturbed sample paths of $\eta(t)$ because of breakages, partial repairs, maintenance interventions in the system, etc.

How can we remedy this situation? When we take a closer look at maintenance practice, we will always discover that maintenance decisions are based on a cleverly chosen *one-dimensional* system (component) "health index." A good car mechanic would take the engine noise as such an index. We would decide to sell our car if the yearly service cost (an integrated parameter!) exceeded some limit. A production line would be stopped for maintenance if the defective rate or a certain product's physical parameter exceeded some critical value. Hastings (1969), Gertsbakh (1972; 1977), Bailey and Mahon (1975), and Kordonsky and Gertsbakh (1995; 1998) present examples of the use of one-dimensional parameters as the basis for making maintenance decisions.

Summing up, it would be desirable to replace the vector $\eta(t)$ by a *scalar* process $r(t) = \psi(\eta_1(t), \ldots, \eta_n(t))$ and to use this new "system health index" as the basis for our maintenance decisions.

5.7.2 The Best Scalarization

The essential feature of the scalar function $r(t)$ must be the ability to discriminate between failed and nonfailed objects. Constructing such a function involves ideas from discriminant analysis, see Anderson (1984, Chap. 6), Gertsbakh (1972; 1977).

Let us assume that the whole population of objects can be conventionally divided into two groups: A (the nonfailed or "good" ones) and B (the failed or "unfit" ones). Such a subdivision cannot always be done easily. Medical doctors can usually distinguish very well between a healthy person and a sick one, and moreover, classify the patient within a specific group according to his/her illness. In technology, however, such a classification is sometimes very difficult. Indeed, between the states "brand new" and "entirely useless" there are many intermediate states which are difficult to distinguish. In practice, however, the decision is made on the basis of intuition, experience and common sense.

In our case, we will need to distinguish between two extreme categories of objects: "very good" and "very bad." Thus, we begin by assuming that there is an expert who can classify all objects under consideration into these two groups, A and B. Let N_A and N_B be the numbers of objects in these two groups, which are characterized by the samples

$$\mathbf{x}_1^A, \dots, \mathbf{x}_{N_A}^A \text{ and } \mathbf{x}_1^B, \dots, \mathbf{x}_{N_B}^B, \tag{5.7.2}$$

where every vector in the sample $z = A, B$, $\mathbf{x}_j^z$, $j = 1, \dots, N_z$, is an n-tuple

$$\mathbf{x}_j^z = (x_{j1}^z, \dots, x_{jn}^z). \tag{5.7.3}$$

The idea of constructing the "best" scalarization is based on replacing the vector $\mathbf{x}_j^z$ by a scalar

$$r_j^z = \sum_{i=1}^{n} l_i x_{ji}^z, \tag{5.7.4}$$

where the coefficients l_i are selected in some "optimal" way.

To explain this approach, let us give a geometric interpretation to the scalarization (5.7.4). The samples A and B are two clusters of points in the n-dimensional parameter space. Let μ_A and μ_B be the n-dimensional vectors of mean values for A and B, respectively, and let $\mathbf{S}_A$ and $\mathbf{S}_B$ be respective sample variance–covariance matrices:

$$\hat{\mu}_A^{(i)} = \sum_{m=1}^{N_A} x_{mi}^A / N_A, \tag{5.7.5}$$

$$\hat{\mu}_B^{(i)} = \sum_{m=1}^{N_B} x_{mi}^B / N_B, \; i = 1, 2, \dots, n, \tag{5.7.6}$$

$$\hat{\mu}_z = (\hat{\mu}_z^{(1)}, \dots, \hat{\mu}_z^{(n)}), \;\; z = A, B\,, \tag{5.7.7}$$

and $\mathbf{S}_z = ||V_{ij}^z||$, $i, j = 1, 2, \ldots, n$, $z = A, B$, where

$$V_{ij}^z = N_z^{-1} \sum_{m=1}^{N_z} (x_{mi}^z - \hat{\mu}_z^{(i)})(x_{mj}^z - \hat{\mu}_z^{(j)}). \tag{5.7.8}$$

Linear transformation means replacing each vector observation $\mathbf{x}_j^z$ by the scalar $\sum_{i=1}^n \ell_i x_{ji}^z$. Geometrically it is equivalent to projecting each observation onto some line whose direction is collinear to the vector $\boldsymbol{\ell} = (l_1, \ldots, l_n)$. Figure 5.10 shows two clusters A and B in a two-dimensinal space. Obviously, the projection onto the direction $\boldsymbol{\ell}^{(1)}$ discriminates better between A and B than does the projection onto $\boldsymbol{\ell}^{(2)}$.

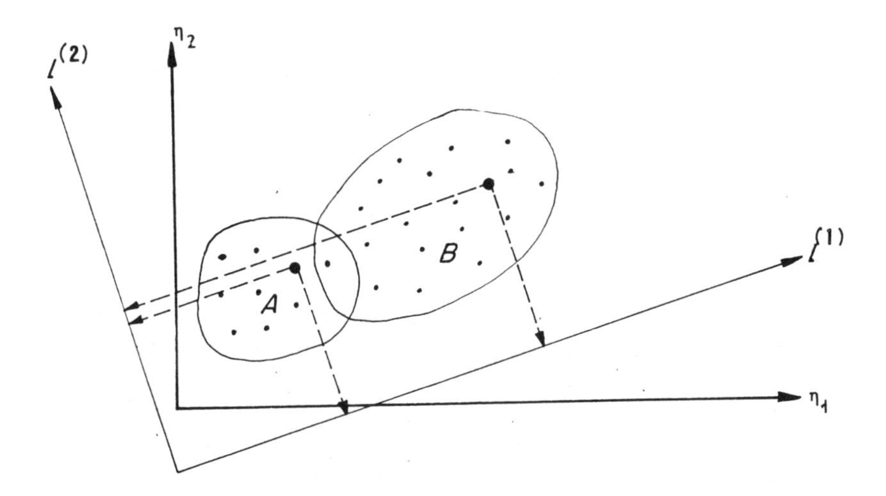

Figure 5.1: Scheme illustrating the choice of the "best" scalarization

The fundamental idea due to R. Fisher is to choose the direction $\boldsymbol{\ell}$ in such a way that the ratio of squares of the difference of projected mean values on the line $\boldsymbol{\ell}$ to the sum of the variances of the projected samples would be *maximal.* Let us represent Fisher's ratio via the vectors of the mean values and the variance–covariance matrices.

Suppose that each vector observation is replaced by the projection

$$y_j^z = l_1 x_{j1}^z + \ldots + l_n x_{jn}^z = \boldsymbol{\ell}\mathbf{x}_j^z, \tag{5.7.9}$$

where $\boldsymbol{\ell}$ is a row vector, $\mathbf{x}_j^z$ is a column vector, and $\boldsymbol{\ell}\mathbf{x}_j^z$ is their scalar product. The mean value in the projected sample z is

$$\hat{y}^z = \sum_{j=1}^{N_z} \sum_{k=1}^{n} x_{jk}^z / N_z = \boldsymbol{\ell}\hat{\mu}_z \tag{5.7.10}$$

($\hat{\mu}_z$ is a column-vector). The variance of the projection of both samples is given by

$$S = \ell\Big(\sum_{z=A,B} (N_z)^{-1} \sum_{j=1}^{N_z} (\mathbf{x}_j^z - \hat{\mu}_z)(\mathbf{x}_j^z - \hat{\mu}_z)' \Big)\ell' = \ell(\mathbf{S}_A + \mathbf{S}_B)\ell'. \quad (5.7.11)$$

The prime denotes transposition. Therefore, Fisher's optimal vector ℓ_* maximizes the following expression:

$$D = \frac{(\ell(\hat{\mu}_A - \hat{\mu}_B))^2}{\ell(\mathbf{S}_A + \mathbf{S}_B)\ell'}. \quad (5.7.12)$$

The following theorem is a well-known fact from linear algebra:

Theorem 5.7.1
Let $\mathbf{S}_A + \mathbf{S}_B$ be a nonsingular matrix. Then D is maximized by the vector ℓ_* defined as

$$\ell_*' = (\mathbf{S}_A + \mathbf{S}_B)^{-1}(\hat{\mu}_A - \hat{\mu}_B). \quad (5.7.13)$$

The maximal value of D equals

$$D^* = (\hat{\mu}_A - \hat{\mu}_B)'(\mathbf{S}_A + \mathbf{S}_B)^{-1}(\hat{\mu}_A - \hat{\mu}_B). \quad (5.7.14)$$

It is instructive to compare the result of this theorem with another approach used in mathematical statistics for the classification of objects into two groups. Let the probabilistic properties of populations A and B be described by their density functions $\mathbf{p}_A(\mathbf{x})$ and $\mathbf{p}_B(\mathbf{x})$, respectively. According to the Neyman–Pearson lemma, the best inference about a given vector $\mathbf{x}_0$'s membership of populations A or B must be formed on the basis of the value of the likelihood ratio

$$r_{A,B}(\mathbf{x}_0) = \mathbf{p}_A(\mathbf{x}_0)/\mathbf{p}_B(\mathbf{x}_0), \quad (5.7.15)$$

or on the value of some other monotone function of $r_{A,B}$.

Suppose that $\mathbf{p}_A$ and $\mathbf{p}_B$ are multidimensional normal densities with identical (nonsingular) covariance matrices $\mathbf{V}$. Then, using standard manipulations, we obtain that

$$\log r_{A,B}(\mathbf{x}_0) = \mathbf{x}_0\mathbf{V}^{-1}(\mu_A - \mu_B) - (\mu_A + \mu_B)'\mathbf{V}^{-1}(\mu_A + \mu_B)/2. \quad (5.7.16)$$

In this formula, the second term is a constant which does not depend on the actual observation vector $\mathbf{x}_0$ and the first term is nothing but a scalar product of the vector $\mathbf{x}_0$ and the vector $\mathbf{V}^{-1}(\mu_A - \mu_B)$. This vector is analogous to the vector ℓ_* in Fisher's approach. Thus there is a close similarity between Fisher and Neyman–Pearson approaches.

5.7.3 Preventive maintenance: multidimensional state description

Let us describe the procedure of data processing and organizing the preventive maintenance for a system whose state description is multidimensional.

Step 1. Two samples representing "brand new" and "entirely unfit" objects are selected. It is desirable that the first sample should consist of well-functioning units operating for a relatively short time. The second sample contains units which have already been in operation for a longer time, and according to expert opinion would be considered unfit or close to failure. The actual values of the parameters have to be measured for every object in the samples, and samples of type (5.7.2) must be formed. It is not expedient to restrict ourselves *a priori* in terms of number of parameters measured (i.e. by the dimension of the vector $\mathbf{x}_j^z$).

Step 2. The vectors $\hat{\mu}_A$, $\hat{\mu}_B$, the matrices $\mathbf{S}_A$, $\mathbf{S}_B$, and $(\mathbf{S}_A + \mathbf{S}_B)^{-1}$ are calculated, and finally the vector ℓ_* defined by (5.7.13). Using this vector, each n-tuple of observations is reduced to a single value y_j^z; see (5.7.9). Now two one-dimensional samples y_j^A, $j = 1, \ldots, N_A$, and y_j^B, $j = 1, \ldots, N_B$, must be investigated in order to decide whether the discrimination between A and B is satisfactory or not. Figure 5.11 shows two examples of comparison of the projected samples.

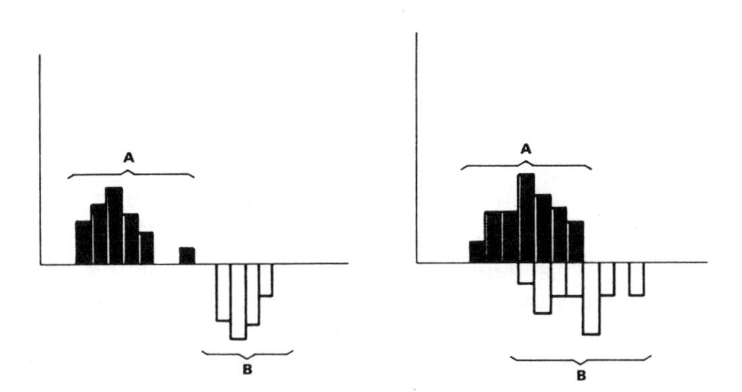

Figure 5.11. Patterns of bad and good discrimination

If the situation is similar to the case shown on the left in Fig. 5.11, we can say that we have succeeded in creating a one-dimensional discrimination function. If not (Fig. 5.11, right), the above discrimination technique cannot be used. We suggest the following half-empirical criterion. Let $\hat{\sigma}_A$ and $\hat{\sigma}_B$ be the standard deviations of the projected samples A and B, respectively. Then we assume that the discrimination is good if

$$|\hat{y}^A - \hat{y}^B| > 2.5(\hat{\sigma}_A + \hat{\sigma}_B) \,. \tag{5.7.17}$$

Step 3. Suppose that we are lucky and there is good discrimination between "good" and "bad" objects. Figure 5.12 shows a possible way of organizing the preventive maintenance. Consider a certain object which at the initial time instant t_0 has vector of parameters $\eta(t_0) = (\eta_1(t_0), \eta_2(t_0), \ldots, \eta_n(t_0))$.

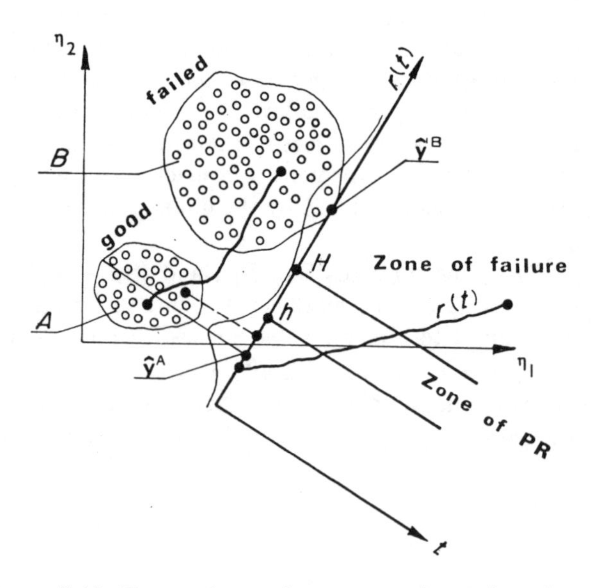

Figure 5.12. Preventive maintenance scheme based on control of the scalar parameter $r(t)$

Until now, when speaking about discrimination, we have acted as if the parameters were "frozen;" their evolution in time has not been taken into consideration. Now recall that $\eta_i(t)$ are random processes. On the parameters' "phase space" shown in Fig. 5.12, the evolution of an object from population A into population B goes along some random path. We are interested only in the projection of the vector $\eta(t)$ on the line ℓ which was chosen as a line with best discriminating properties between A and B. Therefore, our maintenance actions are nothing but a control of the scalar process

$$r(t) = \eta_1(t)l_1 + \eta_2(t)l_2 + \ldots + \eta_n(t)l_n. \tag{5.7.18}$$

This is justified because in the area when this parameter changes, there are clearly expressed zones corresponding to "good" and "bad" objects. Apparently, it is expedient to introduce two levels for $r(t)$, H and h, so that the situation $r > H$ might be considered as a failure, the state $h \le r(t) \le H$ as marginal, and the state $r(t) < h$ as the good one. No maintenance actions should be undertaken if the object is in the good zone; preventive maintenance (PM) must be undertaken if the object is discovered in the marginal zone, and an emergency repair (ER) must be done if the process enters the failure zone. PM and ER mean shifting the process from its actual state to some state in the $(r(t) < h)$ zone.

Step 4. This step is investigation and validation of the stochastic pattern for the behavior of $r(t)$. This is probably the most difficult part of the whole maintenance procedure. First, we will need to observe sample paths of $r(t)$. Very probably, there will be a small number of complete paths and many partial paths since the system under observation is subjected to changes, repairs, etc. We suggest first of all trying to fit a Markov-type process. If discrete levels are introduced for the process $r(t)$, then in fact our first choice is to fit a Markov chain to the stochastic behavior of $r(t)$.

We will not go into the details of this investigation. Let us mention only that Section 3.4 of Gertsbakh (1977) presents an example of such statistical analysis. The objects were hydraulic pumps used in the aircraft industry. A sample of 33 pumps was examined and classified into two groups A and B by an expert, with 25 and 8 units, respectively. Measurements of eight physical characteristics were made, including a pressure oscillation index, spectral characteristics of vibrations and noise. The analysis of the Fisher discrimination ratio showed that for the optimal discrimination, the ratio $(\hat{y}^A - \hat{y}^B)/(\hat{\sigma}_A + \hat{\sigma}_B) = 2.52$. The range of $r(t)$ was divided into 25 levels, where the levels $1, 2, 3$ were defined as "good", the levels $r > 16$ as failure, and the rest as marginal. Statistical analysis based on observing 17 pumps during 3000 hrs of operation allowed a Markov chain model to be constructed with 25 states for $r(t)$. Two inspection periods were introduced: $\Delta = 200$ hrs and $\Delta = 400$ hrs. The inspection, maintenance and ER costs were defined as $c_{insp} = 1, c_{PM} = 10$ and $c_{ER} = 100$, respectively. The optimal policy was found which prescribes inspecting a pump after 400 hrs if it was found in one of the states $1, 2, \ldots, 9$, and inspection every 200 hrs for the states $10, 11, \ldots, 16$. The maintenance action was a shift from marginal or failure state to one of the good states. It was also established that the optimal policy is robust with respect to the changes in cost parameters.

Chapter 6

Best Time Scale for Age Replacement

Counting time is not so important as making time count.
James J. Walker

You will never "find" time for anything. If you want time, you must make it.
Charles Buxton

6.1 Introduction

As a rule, the lifetime of any system (component) can be measured and observed on more than one time scale. There are two widely used "parallel" time scales for cars: the calendar time scale and the mileage scale. For aircraft frame, three time scales are popular: the calendar time scale, the number of flight hours (time in the air) and the number of takeoffs/landings. For jet engines, the age is measured in operation hours and in operation cycles.

In some cases, the choice of the most "relevant" time scale is obvious. For example, the "true" age of a car body must be measured in calendar time, and not in mileage because the main aging factor is corrosion which goes on in "usual" time. On the other hand, the age of the braking system must be measured on the mileage scale since the dominating factor is the wear of brake disks and shoes which develops as a function of mileage. For an aircraft undercarriage, the lifetime must be related to the number of operation cycles (the number of takeoffs and landings).

There are, however, many cases where the right choice of the most relevant time scale is not so obvious. For example, aircraft frame failure is a result of fatigue cracks in the most heavily stressed joints. The appearance of these cracks depends on both corrosion and fatigue damage accumulation. Corrosion developes in calendar time and as a function of time in the air, and fatigue cracks may depend on the total time in air and the number of heavy-stress cycles which take place on takeoffs and landings.

Attempts are being made to design by technical means so-called damage counters which create their own artificial time scale, see e.g. Kordonsky and Gertsbakh (1993).

The correct choice of the "best" time scale is of crucial importance for accurate failure prediction and for planning preventive maintenance activities. Let us consider a very simple but instructive example: replacement of a cutting tool on an automatic production line (APL).

It is known that the milling cutter is capable of providing the necessary accuracy during, say 1000–1200 operation cycles. (Suppose, that the instrument manufacturing process as well as the work of the APL are highly stable well-controlled processes). The APL works for 2–18 hours a day, depending on the market demand, averaging 10 hours. One operation cycle lasts exactly 10 minutes. Thus, on the average, the APL completes 600/10=60 cycles a day. Suppose that the maintenance manager decides to carry out the preventive replacement of the tool every 16 days. On "average", this corresponds to 960 operation cycles, which seems to be a reasonable age. But obviously, this is not a very wise decision: due to large variations in the daily operating time, the tools will actually be replaced every 200 – 1700 cycles. In most cases, either a good new tool will be replaced or one which has already worn out. It would be much better to record the APL's operating time and initiate the replacement as soon as this number reaches the "critical age" of 960 cycles.

In more formal language, there are two time scales for the milling cutter, the operating time and calendar time, and the relevant time scale is the operating time.

Once there are at least two observable principal time scales (say, mileage and operating time), it is always possible to introduce a new time scale by considering a weighted sum of the principal time scales. In this way we arrive at the problem of choosing the best, most appropriate time scale.

Kordonsky suggested the following definition of the "best" time scale (see Kordonsky and Gertsbakh 1993). The best time scale provides the ***minimal coefficient of variation*** (c.v.) of the time to failure. There is a good intuitive reason behind this definition. If on a certain time scale a system has a lifetime whose c.v. is very small, then failure prediction on this scale will be almost certain, and maintenance actions might be carried out very efficiently. We have already demonstrated in Chap. 4 that the c.v. is one of the main factors influencing maintenance efficiency.

In the next section we define formally the the best (optimal) time scale for the family of time scales obtained as a convex combination (with nonnegative weights) of two principal time scales. In Sect. 6.3 we consider an example of fatigue test data and demonstrate the difference in the c.v. between various time scales. We introduce a cost criterion for age replacement for an arbitrary time scale and describe a numerical procedure for finding the optimal time scale for age replacement. In Sect. 6.4 this procedure is applied to the fatigue test data and it is demonstrated that the optimality in the terms of the c.v. and in the terms of the cost criterion for age replacement practically coincide.

Several types of data are considered. For complete samples, we use a nonparametric approach to finding the optimal preventive maintenance policy based on the empirical lifetime data; see also Arunkumar (1972). For censored and quantal-type data, our approach is based on assuming a parametric lifetime distribution model and on using maximum likelihood for parameter estimation.

6.2 Finding the Optimal Time Scale

We need some notation. Time scales are denoted by script letters $\{\mathcal{L},\ \mathcal{T},\ \mathcal{H}\}$. Italic capitals $L,\ T,\ H$ and small italic letters $l,\ t,\ h$ are used to denote random variables and nonrandom variables (constants) in the corresponding time scales, respectively.

As soon as there is more than one scale, the question of which time scale is better arises immediately. Intuitively we prefer that time scale which is able to predict more accurately the failure instant. We suggest the following:

Definition 6.2.1
Suppose that we have several time scales $\mathcal{L}_1,\ \mathcal{L}_2, \ldots, \mathcal{L}_k$, and let the lifetimes of a particular system be $L_1, L_2, \ldots, L_k$, respectively. Then we say that the time scale $\mathcal{L}_1$ is ***optimal*** for this system if the corresponding c.v. is the smallest:

$$c.v.[L_1] = \min_{1 \le i \le k} \{c.v.[L_i]\}. \tag{6.2.1}$$

Recall that the c.v. is defined as

$$c.v.[L] = \sqrt{Var[L]}/E[L]. \tag{6.2.2}$$

The c.v., by (6.2.2), is dimensionless, it is invariant with respect to the choice of time unit and is relatively easy to estimate, especially if a complete sample is available.

When we consider a particular problem for which there is an optimality criterion, the best time scale becomes well defined: it is the time scale which provides the optimal value for this criterion. We examine further an example for which the optimality in the terms of the cost criterion for age replacement and the optimality in the terms of the c.v. practically coincide.

Suppose that we have two principal observable time scales $\mathcal{L}$ and $\mathcal{H}$. Then let us consider the family of times scales

$$\mathcal{T}_a = (1-a)\mathcal{L} + a\mathcal{H}, \quad a \in [0,1]. \tag{6.2.3}$$

This expression means that if the given system has lifetimes L and H on the time scales $\mathcal{L}$ and $\mathcal{H}$, respectively, then the lifetime on the $\mathcal{T}_a$ scale is defined as

$$T_a = (1-a)L + aH \ . \tag{6.2.4}$$

Theorem 6.2.1
Let $\mathbf{S}$ be a nonsingular variance–covariance matrix for the vector $\mathbf{V} = (L, H)$ and let $\mathbf{M}$ be a column vector with components $(E[L], E[H])$. Suppose that both components of the vector $\mathbf{G} = (g_1, g_2)' = \mathbf{S}^{-1}\mathbf{M}$ have the same sign, where the prime means transpose. Then the optimal $a = a^*$ giving the minimum c.v. is

$$a^* = \frac{g_1}{g_1 + g_2} \ . \tag{6.2.5}$$

If g_1 and g_2 have opposite signs, the optimal scale is either $\mathcal{L}$, or $\mathcal{H}$, depending on which has the smaller c.v.

Proof
Finding the minimum of the c.v. is equivalent to maximizing the ratio $(\mathbf{M}'\mathbf{G})^2/\mathbf{G}'\mathbf{S}\mathbf{G}$, $\mathbf{G} \neq \mathbf{0}$. It is known (see e.g. Johnson and Wichern 1988 p. 64), that the maximum is attained at $\mathbf{G} = Const \times \mathbf{S}^{-1}\mathbf{M}$. The minimum value of the c.v. is $1/\mathbf{M}'\mathbf{S}^{-1}\mathbf{M}$. From here it follows that

$$a^* = \frac{g_1}{g_1 + g_2}, \ g_1/g_2 = \frac{E[H]Var[L] - E[L]Cov[L,H]}{E[L]Var[H] - E[H]Cov[L,H]}. \tag{6.2.6}$$

It is very instructive to have a geometric interpretation of the $\mathcal{T}_a$ scale. This scale is obtained by projecting the realizations of a two-dimensional random variable (L, H) on a direction which is collinear to the vector $\mathbf{b} = (1-a, a)$, and by multiplying them by the norm of this vector. (Recall that (6.2.4) is a scalar product of two vectors (L, H) and $(1-a, a)$; see also Fig. 6.1).

6.3 Optimal Time Scale for Age Replacement: Fatigue Data

Example 6.3.1: Kordonsky's fatigue test data
A sample of 30 steel specimens were subjected to cyclic bending until they broke of fatigue. The sample was divided into six groups of size 5, and each group was subjected to a two-level periodic loading consisting of $5000\alpha_j$ low-load cycles and $5000(1-\alpha_j)$ high-load cycles. The lifetimes in terms of the number of low- and high-load cycles are given in the Table 6.1.

Denote by L the number of low-load cycles and by H the number of high-load cycles. The c.v. differ significantly: $c.v.[L] = 1.106$ and $c.v.[H] = 0.399$.

Table 6.1: The number of low-load and high-load cycles to fatigue failure

i	α_j	Low-load	High-load	i	α_j	Low-load	High-load
1	0.95	256 800	13 500	16	0.40	32 000	45 700
2		235 800	11 600	17		48 000	70 400
3		370 150	19 250	18		42 000	61 500
4		335 100	17 500	19		42 000	60 600
5		380 300	20 000	20		54 000	80 400
6	0.80	153 000	38 000	21	0.20	10 000	37 500
7		176 200	44 000	22		16 000	62 700
8		160 300	40 000	23		12 000	45 300
9		156 000	39 000	24		19 000	72 600
10		103 000	25 000	25		11 000	42 000
11	0.60	84 000	54 400	26	0.05	3 000	53 900
12		81 000	52 300	27		3 750	68 550
13		90 000	59 900	28		4 250	77 950
14		57 000	37 300	29		3 320	57 950
15		66 000	42 700	30		2 750	51 250

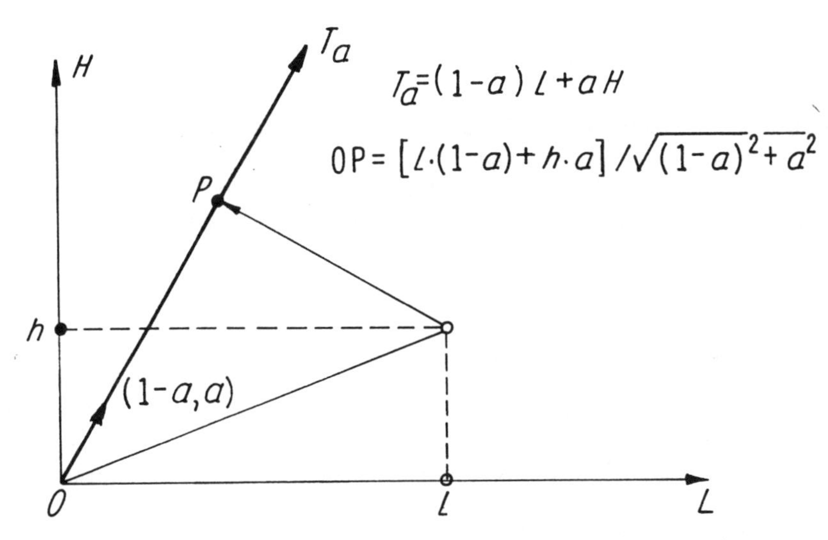

Figure 6.1. The $\mathcal{T}_a$ scale is determined by the vector $(1-a, a)$

Let us proceed with the derivation of a formula for long-run mean costs for age replacement on the $\mathcal{T}_a$ scale. Denote by $F_a(t)$ system lifetime on the time scale (6.2.4). We already know from Chap. 4 the formula for the average long-run costs:

$$\eta_a(x) = \frac{F_a(x) + (1 - F_a(x))d}{\int_0^x (1 - F_a(t))dt} . \tag{6.3.7}$$

(As usual, the cost of failure is 1, the cost of preventive repair is $d < 1$; assume that these costs are measured in dollars).

The optimal preventive maintenance age is the value $x = x_a$ which minimizes $\eta_a(x)$. In (6.3.7), $F_a(t)$ is the system c.d.f. on the time scale $\mathcal{T}_a = (1-a)\mathcal{L} + a\mathcal{H}$, which we call the $\mathcal{T}_a$ scale.

An important issue is comparing costs per unit time on *different* time scales. $\eta_a(x)$ has the dimension of dollars per unit of $\mathcal{T}_a$ time. In order to be able to compare costs for different scales we must express them in the same time unit. We will do it by reducing the dimension of the cost criterion to dollars per unit of $\mathcal{L}$ time.

Suppose e.g. that on the flight time scale the value of $\eta(x)$ is \$100 per hour. Suppose that on the landing time scale the costs are \$300 per landing. To compare these costs we note that on average one landing corresponds, for example, to 4.5 flight hours. Thus \$300/1 landing = \$300/4.5 hours = \$66.67 per one hour. We have used the fact that the average lifetime in hours m_1 and the average lifetime in landings m_2 are related by $m_1 = 4.5m_2$.

In general, let the mean life of the system on scale $\mathcal{L}$ be $E[L]$ and the mean life on the $\mathcal{T}_a$ scale be $E[T_a] = (1-a)E[L] + aE[H]$. Therefore, one unit of $\mathcal{T}_a$ time is equivalent to $E[L]/E[T_a]$ units of $\mathcal{L}$ time. Therefore, to convert the dimension of (6.3.7) into dollars per unit of $\mathcal{L}$ time, we must consider the following expression:

$$\gamma_a(x) = \eta_a(x)E[T_a]/E[L]. \tag{6.3.8}$$

Note that the factor $1/E[L]$ is the same for all a and can be omitted when $\eta_a(x)$ are compared for various a values.

A closer look at (6.3.8) reveals that, up to a factor $1/E[L]$, $\gamma_a(x)$ is nothing but an inverse of the efficiency Q introduced in Sect. 4.3:

$$Q_a(x) = \frac{1}{\eta_a(x)E[T_a]} = \frac{1}{\eta_a(x)\int_0^\infty (1 - F_a(t))dt}. \tag{6.3.9}$$

Suppose now that for fixed a, x_a is the optimal age:

$$\min_{x>0} \gamma_a(x) = \gamma_a(x_a). \tag{6.3.10}$$

Then the optimal $\mathcal{T}_a$ scale corresponds to such $a = a^*$ as *minimizes* $\gamma_a(x_a)$:

$$\gamma_{a^*}(x_{a^*}) = \min_{0 \le a \le 1} \gamma_a(x_a). \tag{6.3.11}$$

Since $\eta_a(x)$ has the dimensions of dollars per unit of $\mathcal{T}_a$ time, and $E[T_a]$ has the dimensions of $\mathcal{T}_a$ time, Q_a is dimensionless with respect to any time scale. Q_a is in fact an "internal" efficiency measure which fits any time scale. This is particularly convenient for comparing various time scales with respect to the efficiency of the age replacement.

6.4 Algorithm for Finding the Optimal $\mathcal{T}_a$

6.4.1 Complete Samples

Nonparametric approach

Let us adopt the *nonparametric* approach according to which the c.d.f $F_a(t)$ in (6.3.7) is replaced by the empirical distribution function $\hat{F}_a(t)$. Let $l_1, \ldots, l_m$ and $h_1, \ldots, h_m$ be the observed lifetimes on the $\mathcal{L}$ and $\mathcal{H}$ time scales, respectively. Then for fixed a we have a complete sample of $\mathcal{T}_a$ times: $t_i = (1-a)l_i + ah_i$, $i = 1, 2, \ldots, m$. Assume that t_i are already ordered: $t_1 < t_2 < \ldots < t_m$. Then the empirical distribution function is (see Sect. 3.2):

$$\hat{F}_a(t) = \frac{i}{m}, \; t \in [t_i, t_{i+1}), \; i = 1, \ldots, m-1, \tag{6.4.12}$$

with $\hat{F}_a(t) = 0$ for $t < t_1$, and $\hat{F}_a(t) = 1$ for $t \geq t_m$. The nonparametric estimate of (6.3.7) is

$$\hat{\gamma}_a(x) = \frac{[\hat{F}_a(x) + (1 - \hat{F}_a(x))d] \cdot \int_0^{t_m} (1 - \hat{F}_a(t))dt}{\int_0^x (1 - \hat{F}_a(t))dt} (E[L])^{-1}. \tag{6.4.13}$$

The procedure for finding the optimal scale is now as follows. Fix $a = a_0$; find the optimal PM age x_0 which minimizes $\hat{\gamma}_{a_0}(x)$. Denote this minimum value by $\hat{\gamma}^*_{a_0}$; repeat for $a := a + \Delta a$ on a grid over $a \in [0, 1]$. Find that a^* which minimizes $\hat{\gamma}^*_{a_0}$:

$$\gamma^* = \min_{0 \leq a_0 \leq 1} \hat{\gamma}^*_{a_0}. \tag{6.4.14}$$

It follows from the general theory that as the sample size $m \to \infty$, the optimal PM age tends for fixed a with probability 1 to the true optimal PM age on the $\mathcal{T}_a$ scale.

Parametric approach

Let us assume that the lifetime T_a has a known parametric form $T_a \sim F(v; \theta(a))$, where $F(\cdot)$ has a known functional form, and $\theta(a)$ is an unknown parameter, possibly multidimensional. Denote by $f(v; \theta(a))$ the corresponding density function. Our goal is to estimate $\theta(a)$ and to find the "best" value of a. We suggest using a method based on maximum likelihood, which works as follows.
For each fixed a, write the likelihood function for the observations $\{(l_i, h_i),$ $i = 1, \ldots, m\}$:

$$Lik = \prod_{i=1}^{m} f((1-a)l_i + ah_i; \theta(a)). \tag{6.4.15}$$

Maximize the log-likelihood with respect to $\theta(a)$. Let $\hat{\theta}(a)$ be the MLEs.

Express the coefficient of variation of T_a via $\hat{\theta}(a)$:

$$c.v._{(a)} = \psi(\hat{\theta}(a)). \tag{6.4.16}$$

Put $a := 0.00(0.01)1.00$. Denote by $a = a^*$ that value which minimizes the c.v. Then the "best" time scale will be $\mathcal{T}_{a^*}$, and the corresponding c.d.f. will be $F(v; \hat{\theta}(a^*))$.

6.4.2 Right-Censored Samples

A procedure similar to the above one for complete samples can be extended to the case of right-censored L and H times. Suppose that the ith item was observed until time l_i on the $\mathcal{L}$ scale and was removed from the observation afterwards. Suppose that this item survived time h_i on the $\mathcal{H}$ scale. Then obviously the lifetime on $\mathcal{T}_a$ is right-censored by $t_a(i) = (1-a)l_i + ah_i$.

The key idea is to calculate the Kaplan–Meier estimate of the survival function $\hat{R}_a(t)$ and to use $1 - \hat{R}_a(t)$ instead of the empirical distribution function $\hat{F}_a(t)$ in (6.4.13).

There is, however, a complication: for censored data, the Kaplan–Meier estimate is available usually only for a finite interval $[0, T_{max}]$, where T_{max} depends on the largest noncensored observation; see Sect. 3.1. Typically, this interval is large enough for finding the optimal replacement age, but it is not clear how to estimate the mean lifetime on $\mathcal{T}_a$. For this purpose we recommend the use of a "parametric assumption:" plot the data on probability paper, estimate approximately the parameters from the plot which seems to produce the best fit, estimate $E[T_a]$ and plug this estimate into (6.4.13).

Parametric approach

Here the situation is much easier and recalls the case of complete samples, the only difference being the form of the likelihood function:

$$Lik = \prod_{i=1}^{r} f((1-a)l_i + ah_i; \theta(a)) \prod_{i=r+1}^{m} \Big(1 - F((1-a)l_i + ah_i; \theta(a))\Big). \tag{6.4.17}$$

Here $i = 1, \ldots, r$ are the numbers of complete observations, and $i = r+1, \ldots, m$ the numbers of the right-censored ones. The computation now proceeds in a manner similar to the procedure in the previous subsection.

6.4.3 Quantal-Type Data: Parametric Approach

In real-life situations, very often the reliability data come from periodic inspections and thus there are no "complete" observations at all. Below we will consider an example in which the data are based on a single inspection.

Suppose that an item is examined at some point (l_i, h_i) in the (L, H) plane. The information received is as follows: failure exists or does not exist. Geometrically, if the failure exists, its appearance point has coordinates within the rectangle with vertices $(0,0),(0,h_i),(l_i,0),(l_i,h_i)$. On the $\mathcal{T}_a$ time scale this means that the failure appeared *before* the instant $t_a(i)=(1-a)l_i+ah_i$. We are in a situation described in Sect. 3.1.1 as "Quantal response data." Accordingly, the likelihood function is

$$Lik=\prod_{i=1}^{r}F((1-a)l_i+ah_i;\theta(a))\prod_{i=r+1}^{n}(1-F((1-a)l_i+ah_i;\theta(a))),\qquad(6.4.18)$$

where the items with the "yes" response are numbered $i=1,\ldots,r$ and the items with the "no" response are numbered $i=r+1,\ldots,m$.

The search for the "best" time scale goes exactly as described above, with the obvious change in the likelihood function. The next section presents two examples based on fatigue data.

6.5 Optimal Age Replacement for Fatigue Test Data

Example 6.3.1 continued

The data of Example 6.3.1 are plotted on Fig. 6.2 The algorithm described in Section 6.4.1 was applied with a changing on the grid 0.0(0.02)1.0. Table 6.2 shows the optimal age replacement parameters.

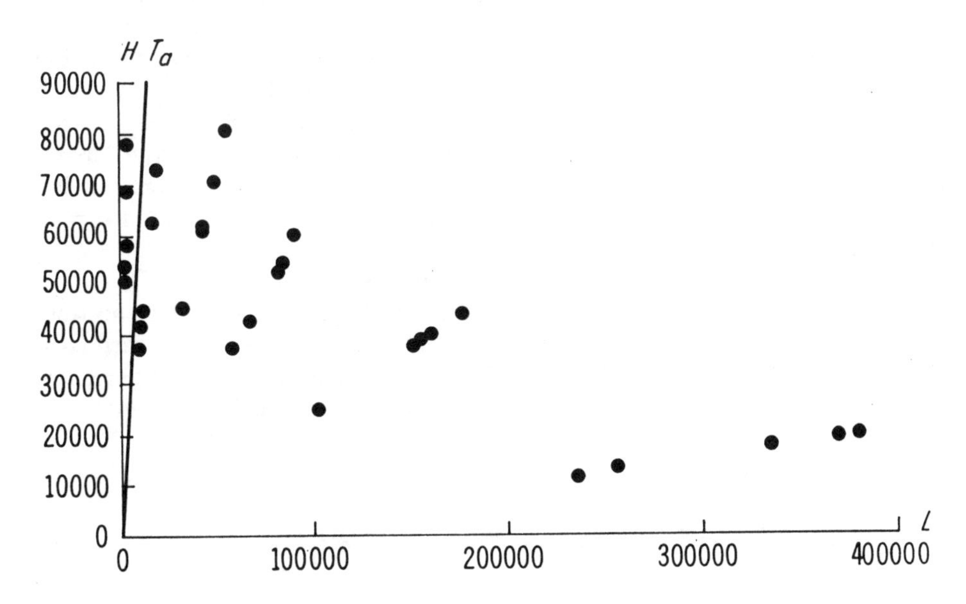

Figure 6.2. The fatigue test results on the (L,H) plane

Table 6.2: Optimal age replacement parameters

d	Optimal a	Optimal age x_a	r	Efficiency Q
0.1	0.88	34,200	1	5.60
0.2	0.88	34 200	1	3.01
0.3	0.88	34 200	1	2.06
0.4	0.88	38 280	3	1.59
0.5	0.88	38 280	3	1.32

Table 6.3: Efficiency Q as a function of a

a	0	0.4	0.6	0.8	0.88	1.0
Q	1.03	1.04	1.29	1.86	2.06	1.58

For small $d = 0.1$–0.3, the optimal replacement age coincides with the smallest observation t_1. It increases as d increases, which is typical for age replacement. For $d = 0.4$–0.5, the optimal age coincides with the third ordered observation t_3. For all values of d the optimal $a = 0.88$. Most interesting is the fact that this a value is very close to $a^* = 0.873$ which provides the minimal c.v.! Note that the minimum value of the c.v. equals 0.399, which is considerably smaller than the c.v. in the $\mathcal{L}$ or $\mathcal{H}$ scale. The last column shows that the optimal age replacement provides a considerable increase in efficiency.

How strong is the influence of the parameter a, i.e. the choice of the time scale, on the efficiency of the age replacement? Table 6.3 gives the answer for $d = 0.3$. It is seen that the optimal scale with $a = 0.88$ is considerably more efficient than the best of the two scales $\mathcal{L}$ or $\mathcal{H}$ with $a = 0$ and $a = 1$, respectively.

Example 6.5.1: Fatigue cracks in aircraft wing joints

Table 6.4 shows the results of inspecting aircraft wing joints. At each inspection, the *size* of the fatigue crack Δ_R and Δ_L in millimeters has been recorded (Kordonsky and Gertsbakh 1997). (L and R indicate the left and the right joint, respectively). Find the optimal time scale and the optimal age replacement policy.

The likelihood function is

$$Lik = \prod_{i=1}^{34} [F((1-a)l_i + ah_i; \theta(a))]^{\delta_i} [1 - F((1-a)l_i + ah_i; \theta(a))]^{1-\delta_i}, \quad (6.5.19)$$

where $\delta_i = 1$ if the object i had a crack, and $\delta_i = 0$, otherwise.

Our principal assumption is that the lifetime on the $\mathcal{T}_a$ scale has a lognormal

Table 6.4: Cracks in wing joints

Object i	Flight time L_i $[hrs]$	# flights H_i	Δ_R	Δ_L
$1, 2$	30	15	0	0
$3, 4$	40	20	0	0
$5, 6$	60	40	0	0
$7, 8$	80	60	0	0
$9, 10$	170	129	0	0
$11, 12$	180	128	0	0
$13, 14$	230	140	0	0
$15, 16$	250	140	0	0
$17, 18$	350	170	0	0
$19, 20$	380	160	0	0
$21, 22$	640	290	0	0
$23, 24$	670	320	0	0
$25, 26$	780	260	4	0
$27, 28$	840	420	0	9
$29, 30$	860	430	3	5
$31, 32$	560	470	3	4
$33, 34$	960	360	0	5

distribution:

$$P(\tau_a \leq x; \mu_a, \sigma_a) = \Phi\Big(\frac{\log x - \mu_a}{\sigma_a}\Big). \tag{6.5.20}$$

For each a on a discrete grid 0.00(0.01)1.00, we found the corresponding MLEs of μ_a and σ_a. For the lognormal distribution, the c.v. depends only on σ_a; see Sect. 2.3. The minimal c.v. and therefore the "best" scale correspond to the *minimal* σ_a. Below are the results:

For the $\mathcal{L}$-scale: $a = 0$, $\mu_a = 6.64$. The standard deviation $\sigma_a = 0.463$.

For the $\mathcal{H}$-scale: $a = 1$, $\mu_a = 5.87$. The standard deviation $\sigma_a = 0.262$.

For the "best" scale: $a = 0.852$, $\mu_a = 6.036$. The standard deviation $\sigma_a = 0.197$.

The best time scale is $\mathcal{T}_{0.852} = 0.148\mathcal{L} + 0.852\mathcal{H}$.

Let $d = 0.2$, which is a rather high cost for inspection and PM.

On the "best" scale, the optimal age for the PM is $T^*_{0.852} = 302$. The efficiency of the optimal age replacement $Q = 1.6$.

Let us check the probability of crack appearance on the interval $[0, 302]$. It equals $\Phi((\log 302 - 6.036)/0.197) = \Phi(-1.65) \approx 0.05$.

Suppose that it is necessary to guarantee a much smaller failure probability $p = 0.001$. Then the maintenance period x must satisfy the equation $p = 0.001 = \Phi((\log x - 6.036)/0.197)$, or $x \approx 230$. Figure 6.3 shows the plot of

$\eta_{a^*}(T)$. $T = 230$ will reduce the efficiency of the optimal age replacement to $Q = 1.45$.

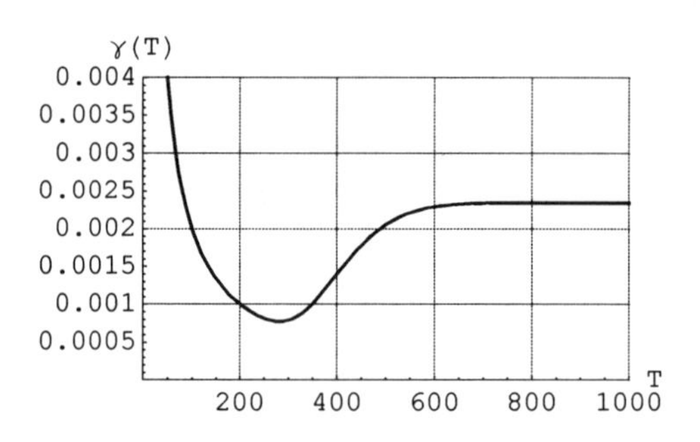

Figure 6.3. $\eta_{a^*}(T)$ for Example 6.5.1

Chapter 7

Preventive Maintenance with Learning

What is all Knowledge too but recorded Experience, and a product of History; of which, therefore, Reasoning and Belief, no less than Action and Passion, are essential materials ?

Carlyle, *Essays*

Learned fools are the greatest fools.

Proverb

7.1 Information Update: Learning from New Data

An important feature has been missing in all preventive maintenance models considered so far. This feature is *learning.* Our decisions regarding the timing and type of maintenance actions were based solely on the information about the system, which we had *before* we actually started implementing the maintenance policy.

Denote the information we initially had by *Data(0).* Then our policy so far was the following: *The decisions on preventive maintenance actions during the whole system life period were determined solely on the basis of Data(0).*

In reality, during system operation there is always an information influx regarding system properties, maintenance costs, etc. The main sources of this additional information are new laboratory test data, field data coming from similar systems, new information from the producer, observations on the system under service, etc.

The right thing to do is to *reconsider*, in the light of this new knowledge, the decisions which were taken previously and were based on lesser knowledge. So we arrive at preventive maintenance with *learning*.

The formal framework for preventive maintenance with learning will be the so-called "rolling horizon" model. We consider the policy of inspection/renewal of a system based on knowledge of its lifetime distribution. Our initial knowledge denoted $Data(0)$ leads to the following decision: carry out inspections of the system at the instants $\{t_1, t_2, t_3, \ldots\}$. This sequence maximizes the reward for the given $Data(0)$. We follow this policy only one step, i.e. during the period $[0, t_1]$. At the instant t_1, we reconsider all information we have: $Data(0)$ plus all new information received during $[0, t_1]$. This constitutes our new information denoted symbolically as $Data(1) = Data(0) \odot data(1)$. Now we look for the optimal maintenance policy for the remaining time (starting at t_1), given $Data(1)$. Suppose that the "best" inspection times now are $\{t_2^*, t_3^*, \ldots\}$. Carry out the first inspection at time t_2^*, reconsider at t_2^* the inspection policy in the light of new $data(2)$, check the system at the next optimal inspection instant, update the information, etc.

The information regarding system lifetime consists of two parts: permanent and variable. The permanent part is the list of all feasible lifetime c.d.f.'s:

$$\{F_1(t), F_2(t), \ldots, F_m(t)\}. \tag{7.1.1}$$

The variable part is the collection of prior/posterior probabilities:

$$\{p_1, p_2, \ldots, p_m\}, \ \sum_{i=1}^{m} p_i = 1. \tag{7.1.2}$$

These are subject to change, according to Bayes' theorem, depending on the information received during the service and maintenance process of the system.

Suppose that at a certain stage of this process the variable information (7.1.2) is $\{\hat{p}_1, \hat{p}_2, \ldots, \hat{p}_m\}$. Then our immediate decision regarding the system lifetime will be based on the assumption that the "true" lifetime has c.d.f.

$$\hat{F}(t) = \sum_{i=1}^{m} \hat{p}_i F_i(t), \tag{7.1.3}$$

and that the corresponding lifetime density function is

$$\hat{f}(t) = \sum_{i=1}^{m} \hat{p}_i f_i(t), \tag{7.1.4}$$

where $f_i(t)$ is the density corresponding to $F_i(t)$.

Now let us describe the mechanism of updating the variable part of the information (7.1.2).

We assume that before the system started operating, our initial (prior) information is summarized in the form of the prior distribution:

$$Data(0) = \{p_i^0,\ i = 1, \ldots, m\}\ . \tag{7.1.5}$$

Suppose that at a certain instant we receive *new* information in the form of complete or censored observations regarding system lifetime. For example, we receive the following laboratory test data: two system lifetimes x_1 and x_2 were observed; three other systems did not fail during the testing period $[0, x_3]$.

In order to simplify and make tractable the formal part of our investigation, we postulate the following two properties of the information received in the course of system operation:

(I): The information influx does not depend on the properties of the system under service. In other words, it comes from "external" sources.

Denote by $data(k)$, $k = 1, 2, 3, \ldots$, the data we receive in the course of system operation at time instants t_k, where $t_1 < t_2 < t_3 \ldots$.

(II): $\{data(k),\ k = 1, 2, 3, \ldots\}$ consist of observed values of statistically *independent* random variables.

Let the likelihood function for $data(k)$ be $Lik[data(k)|j]$, if it is assumed that the data are generated by $F_j(t)$. (Recall the material of Sect. 3.3 on the likelihood function).

Now we apply Bayes' theorem to recompute our prior distribution $\{p_i^0,\ i = 1, \ldots, m\}$ into the posterior distribution on the basis of new data influx. After observing $data(1)$, our posterior distribution will be

$$p_j^1 = \frac{p_j^0 \cdot Lik[data(1)|j]}{\sum_{r=1}^m p_r^0 \cdot Lik[data(1)|r]}\ ,\quad j = 1, 2, \ldots, m. \tag{7.1.6}$$

Symbolically, $Data(1) = \{p_j^1,\ j = 1, 2, \ldots, m\}$. The posterior lifetime c.d.f. after observing $data(1)$ will be

$$F^1(t) = F(t|data(1)) = \sum_{j=1}^m p_j^1 F_j(t), \tag{7.1.7}$$

with corresponding density function

$$f^1(t) \equiv f(t|data(1)) = \sum_{j=1}^m p_j^1 f_j(t). \tag{7.1.8}$$

Suppose the variable information was updated k times. Denote the corresponding posterior distribution by $Data(k) = \{p_j^k,\ j = 1, \ldots, m\}$. After receiving the next portion of information $data(k+1)$, the posterior distribution is updated as follows:

$$p_j^{k+1} = \frac{p_j^k Lik[data(k+1)|j]}{\sum_{r=1}^m p_r^k Lik[data(k+1)|r]}\ ,\quad j = 1, 2, \ldots, m. \tag{7.1.9}$$

If our posterior distribution is $\{p_j^{k+1},\ j = 1,2,\ldots,m\}$, we act as if the system lifetime has c.d.f.

$$F^{k+1}(t) = \sum_{i=1}^{m} p_i^{k+1} F_i(t) . \tag{7.1.10}$$

It follows from more advanced statistical theory that if all data come from the "true" c.d.f. $F_\alpha(t)$, and if the amount of information goes to infinity, the sequence of vectors $\{p_j^k,\ j = 1,\ldots,m\}$ converges to the vector $(0,\ldots,1_\alpha,\ldots,0)$ as $k \to \infty$ (see e.g. Kullback 1959).

7.2 Maintenance–Inspection Model with No Information Update

Suppose that, according to our present information, the system lifetime density function is $f(t)$. We assume that a reward of c_1 is accrued for each unit of failure-free operating time, and a penalty c_2 is paid for each unit of operating time after a failure has appeared.

At $t = t_1$ the system is inspected and its true state (failed or not) is revealed. If the inspection reveals failure, the system is repaired and brought to the "brand new" state. The cost of that action is c_f. If the inspection reveals no failure, the system is preventively repaired and also brought to the "brand new" state, at a smaller cost c_p.

All rewards received at time t_1 are discounted by a factor $(1-\beta)^{t_1}$.

Suppose that after the inspection no new information has been received. Denote by $V(f;t_1)$ the mean discounted reward for $[0,\infty]$ if the inspection period is t_1 and the lifetime has d.f. $f(t)$. Following the above description,

$$V(f;t_1) = R(f;t_1) + (1-\beta)^{t_1} V(f;t_1), \tag{7.2.1}$$

where $R(f;t_1)$ is the one-step reward

$$R(f;t_1) = c_1 \int_0^{t_1} x f(x)dx + c_1 t_1 (1 - F(t_1))$$

$$-c_2 \int_0^{t_1} (t_1 - x) f(x)dx - c_f F(t_1) - c_p(1 - F(t_1)) . \tag{7.2.2}$$

Note that in order to simplify the formal treatment of our problem we have adopted a simplified discounting scheme: the rewards received during the first inspection period $[0,t_1]$ are not discounted. The discounting in a form of a factor $(1-\beta)^{t_1}$ is applied only to *future* rewards.

In order to find the best time for the first inspection, let us introduce the maximum discounted reward $V(f)$. This satisfies the following dynamic programming-type recurrent relationship:

$$V(f) = \max_{t>0}[R(f;t) + (1-\beta)^t V(f)]. \tag{7.2.3}$$

Indeed, if we choose the inspection period t, our reward will be $R(f;t)$ for the first step plus the maximum future rewards $V(f)$ discounted by the factor $(1-\beta)^t$.

Suppose that the maximum in (7.2.3) is attained at $t = t^*$. Then the optimal sequence of inspection times is $\{kt^*,\ k = 1, 2, \ldots\}$. $V(f)$ satisfies the equation $V(f) = R(f;t^*) + (1-\beta)^{t^*} V(f)$. Thus $V(f) = R(f;t)/(1-(1-\beta)^t)$ and, by the definition of $V(f)$,

$$t^* = \arg\max_{t>0} \frac{R(f;t)}{1-(1-\beta)^t}\,. \tag{7.2.4}$$

7.3 Rolling Horizon and Information Update

The optimal inspection policy with no information update is periodic, with the period t^* found above. Now we implement information updating for the nomination of the inspection times.

We start our process with $Data(0) = \{p_i^0,\ i = 1, \ldots, m\}$, where p_i^0 is the prior probability that the true lifetime density is $f_i(t)$. We plan the first inspection at t_1^*:

$$t_1^* = \arg\max_{t>0} \frac{R(f^0;t)}{1-(1-\beta)^t}, \tag{7.3.1}$$

where $f^0(t)$ is determined by (7.1.5).

At $t = t_1^*$, new information in the form of $data(1)$ is received. We update our prior information regarding the lifetime density via the Bayes rule (7.1.6). Now we plan the next inspection as if the true lifetime density (in the light of our new knowledge) is

$$f^1(t) = \sum_{j=1}^{m} p_j^1 f_j(t). \tag{7.3.2}$$

Suppose that the second inspection takes place at the instant $t_2 = t_1^* + t_2^*$, where t_2^* is found from the relationship

$$t_2^* = \arg\max_{t>0} \frac{R(f^1;t)}{1-(1-\beta)^t}. \tag{7.3.3}$$

Update again the information, and proceed in a similar way.

A typical $(k+1)$th step of the above procedure is as follows. Suppose that the $(k+1)$th inspection is carried out at the instant $t_{k+1} = t_1^* + \ldots + t_k^* + t_{k+1}^*$.

The information $data(k+1)$ is received at t_{k+1}. Then update the information following the Bayes formula (7.1.9); schedule the next inspection at $t_{k+2} = t_{k+1} + t_{k+2}^*$, where t_{k+2}^* is given by

$$t_{k+2}^* = \arg\max_{t>0} \frac{R(f^{(k+1)};t)}{1-(1-\beta)^t}. \tag{7.3.4}$$

Here

$$f^{(k+1)} = \sum_{j=1}^{m} p_j^{k+1} f_j(t). \tag{7.3.5}$$

7.4 Actual and "Ideal" Rewards

Suppose that the "true" c.d.f. is $F_\alpha(t)$, where α can in principle be any one of the indices $1, 2, \ldots, m$. Let us derive an expression for the actual total reward.

Suppose that using the rolling horizon policy we derived the following sequence of inspection times: $t_1 = t_1^*$, $t_2 = t_1^* + t_2^*, \ldots, t_k = t_1^* + \ldots + t_k^*, \ldots$ Then the total accumulated *actual* reward is

$$V_\alpha = R(f_\alpha; t_1^*) + (1-\beta)^{t_1} R(f_\alpha; t_2^*) + (1-\beta)^{t_2} R(f_\alpha; t_3^*) + \ldots, \tag{7.4.1}$$

where $R(f_\alpha; t_1^*)$ is the one-step mean reward on $[0, t_1^*]$ computed for the lifetime density $f_\alpha(\cdot)$.

It is interesting to compare the actual reward with the maximal reward $V_{\max}$ which would have been received if already at $t = 0$ the true CDF F_α had been known. Obviously,

$$V_{\max} = \max_{t>0} [R(f_\alpha; t) + (1-\beta)^t V_{\max}] \,. \tag{7.4.2}$$

Let us to characterize the relative efficiency of the rolling horizon policy by the ratio

$$\eta = \frac{V_\alpha}{V_{\max}} \,. \tag{7.4.3}$$

7.5 Numerical Example

Our initial information regarding the lifetime is the following : $F(t)$ may be one of the following three c.d.f.s:

$F_1(x) = 1 - e^{-x}$, the exponential distribution;
$F_2(x) = 1 - e^{-x^2}$, the Weibull distribution with $\lambda = 1, \beta = 2$;
$F_3(x) = 1 - e^{-0.25x^2}$, the Weibull distribution with $\lambda = 0.5, \beta = 2$.

Note that F_1 has mean value 1, F_2 has mean 0.886 and F_3 has mean 1.77. The respective density functions are $f_1(x) = e^{-x}$, $f_2(x) = 2xe^{-x^2}$, $f_3(x) = 0.5xe^{-0.25x^2}$.

Our *Data*(0) now is summarized in the following vector of prior probabilities: *Data*(0) $= \mathbf{p}^0 = (1/3;\ 1/3;\ 1/3)$. We assign, therefore, equal probabilities to any one of the above three possibilities.

The reward (cost) parameters are: $c_1 = 50$; $c_2 = 10$, $c_f = 20$, $c_p = 1$. We assume that the discount factor is $\beta = 0.1$.

We start with the lifetime density which is equal to

$$f^0(t) = \sum_{i=1}^{3} p_i^0 f_i(t) = \frac{1}{3}\left(e^{-t} + 2te^{-t^2} + 0.5te^{-0.25t^2}\right). \tag{7.5.1}$$

The relationship (7.2.4) gives $t_1^* = 0.256$.

Suppose that no new information has been received at the instant $t_1 = t_1^* = 0.256$. According to our approach, it has been decided to carry out the next inspection at $t_2 = 0.256 + 0.256 = 0.512$. Again, there was no new information.

New data became available only at the instant $t_4 = 4 \times 0.256 = 1.024$. So $data(1), data(2), data(3)$ were zero. $data(4)$ has the form of three observed lifetimes: $data(4) = \{x_1 = 0.904;\ x_2 = 1.222;\ x_3 = 0.921\}$. These are complete observations. The expression for the likelihood is

$$Lik[data(4)|j] = \prod_{i=1}^{3} f_j(x_i). \tag{7.5.2}$$

Using the Bayes formula (7.1.6) we recompute our prior probabilities and obtain the updated information: $\mathbf{p}^4 = \{0.105;\ 0.767;\ 0.128\}$. We write a superscript "4" since this vector is obtained at t_4.

Now our starting point is the lifetime density
$f^4(t) = \sum_{i=1}^{3} p_i^4 f_i(t) = \left(0.105 \cdot e^{-t} + 0.767 \cdot 2te^{-t^2} + 0.128 \cdot 0.5te^{-0.25t^2}\right)$.
Using (7.2.4) we get $t_5^* = 0.219$. At $t_5 = 1.024 + 0.219 = 1.243$ no new data arrived. We apply the inspection period 0.219 once again, and at time $t_6 = 1.243 + 0.219 = 1.462$ we have a new portion of information which consists of three noncensored observations: $data(6) = \{x_1 = 0.971;\ x_2 = 1.149;\ x_3 = 0.246\}$. Using the Bayes formula we obtain
$\mathbf{p}^6 = \{0.056;\ 0.931;\ 0.013\}$.
It is decided now that the "true" c.d.f. is F_2, i.e. $W(\lambda = 1, \beta = 2)$. We act further as if the prior probabilities were $\mathbf{p} = (0;\ 1;\ 0)$.

The data given above were randomly generated from the population F_2. There is ≈ 0.07 probability that our choice of the "true" lifetime is wrong.

If the true c.d.f. were F_2 then (7.2.4) would provide the answer $t^* = 0.205$. So, on the interval $[t_6,\ \infty]$, the inspection period is constant and equals 0.205.

Suppose that we knew from the very beginning that the true lifetime is F_2. Then we would use this period from the very beginning and by (7.2.3) the mean reward would have been

$$V_{max} = V(f_2) = \max_{t>0}[R(f_2;t) + 0.9^t V(f_2;t)] = 388.3.$$

Let us compute by (7.4.1) the mean *actual* reward. The actual reward corresponds to the "true" lifetime F_2 and the actual nonoptimal choice of the inspection periods. Calculating $R(f_2;t)$ by (7.2.2) for $t = 0.256$, $t = 0.219$, $t = 0.205$, we obtain: $R(f_2; 0.256) = 10.25; R(f_2; 0.219) = 8.36; R(f_2; 0.205) = 8.13$. Calculations by (7.4.1) give $V_\alpha = V_2 = 380.2$. This number is surprisingly close to $V_{\max}$! By (7.4.3), $\eta = 0.97$.

7.6 Exercises to Chapters 5–7

1. A system has three states denoted 1, 2 and 3. The transitions between these states occur according to a Markov chain with transition probabilities $P_{12} = 0.3$, $P_{13} = 0.7$, $P_{21} = 0.1$, $P_{23} = 0.9$, $P_{32} = 1$. The system stays in states 1 and 3 exactly one hour, after which a transition occurs. In state 2 the system stays a random time $\tau \sim \text{Exp}(\lambda = 0.5)$. Each transition gives a reward of \$1. Find the mean reward per unit time for the above system.

2a. Consider a multiline system as described in Sect. 5.2. The time to failure τ of a line has an arbitrary c.d.f. $F(t)$ and density function $f(t)$. Assume that for each line a unit of operational time gives a reward c_{rew} and each unit of idle time gives a negative reward $-c_f$. The system is totally renewed for a cost c_{rep} after k lines out of n have failed. Derive an expression for the reward per unit time.
Hint. Derive first the d.f.s of the order statistics $\tau_{(1)}, \tau_{(2)}, \ldots, \tau_{(n)}$.

2b. Assume that $n = 4$, the lifetime of a line has the c.d.f. $W(\lambda = 0.25, \beta = 1.5)$, $c_{rew} = 1$, $c_f = 0.1$ and $c_{rep} = 2$. Find numerically the optimal value of k.

3. Ten hydraulic high pressure hoses were installed on operating aircraft. The hoses were periodically inspected and deemed to have failed when a microcrack has been detected. The data are as follows:

i	hours to failure	flights to failure
1	3700	920
2	3900	1020
3	4200	1200
4	2700	1370
5	3100	1540
6	1950	1520
7	2100	1630
8	2300	1660
9	2700	1760
10	2800	1840

Consider nonparametric age replacement on the time scales $\mathcal{H}$ (time in hours), $\mathcal{F}$ (number of flights) and in the "best" combined scale $\mathcal{T}_a = (1 - a)\mathcal{H} + a\mathcal{F}$, which provides the smallest value of the c.v. Take $a = 0.0(0.05)1.00$. Assume that ER costs $c_{ER} = 1$, and that PM costs $d = 0.2$.

Compare the costs of the optimal age replacement on the $\mathcal{H}$ and $\mathcal{F}$ scales and on the best combined scale.

4. Consider the following modification of age replacement: replace at age t_p which is the p quantile of c.d.f. $F(t)$. Write a formula for cost per unit time.

Try to simplify it by assuming that $1-F(t)$ changes linearly between the points $[0, t_p]$.

5. *Optimal control of spare parts*
A system has one main part and $r-1$ spare parts. The system has to work during the time period $[0, m+1]$. Any part which is put to work for a unit time period has a constant probability of failure p during this period. Once the part has failed it is never recovered. At any instant $t = k$ it is possible to put to work any number of parts available. For example, if at the instant k there are 5 nonfailed units, then any number i, $1 \leq i \leq 5$ can be put to work for the period $[k, k+1]$. If all parts put to work for one time period $[s, s+1]$ have failed, is deemed to have occured system failure. Find the optimal policy for putting the spare parts to work which would *minimize* the probability of system failure for the period $[0, m+1]$.
Hint. Suppose that k_j parts are put to work at the instant $t = k$ for a period $[k, k+1]$. Assign a unit cost if all these parts have failed and zero cost if the number of failed parts was z, $z < k_j$. Denote by $W_i(m)$ the minimum average cost for a system which has to work m units of time and has initially i nonfailed parts. Write a dynamic programming-type recurrence relation expressing $W_j(m+1)$ through $W_j(m)$. Argue that $W_j(m)$ is equal to the probability of system failure on $[0, m]$. For more details see Gertsbakh (1977, p 147).

6. Suppose that p_j^0 is the prior probability that the "true" density is $f_j(t)$, $j = 1, 2, \ldots, m$. Suppose that the following data, in the form of a complete sample, has been received: $data = \{x_1, \ldots, x_{n-1}, x_n\}$. Then the posterior probabilities are, according to (7.1.6),

$$p_j^1 = \frac{p_j^0 \prod_{j=1}^n f_j(x_i)}{\sum_{k=1}^m p_k^0 \prod_{i=1}^n f_k(x_i)},$$

$j = 1, 2, \ldots, m.$

Suppose that the data come in portions. The first portion is $data(1) = \{x_1, \ldots, x_{n-1}\}$ and the second $data(2) = \{x_n\}$. Then we can act follows. First, compute the posterior probabilities from $data(1)$:

$$\hat{p}_j^1 = \frac{p_j^0 \prod_{i=1}^{n-1} f_j(x_i)}{\sum_{k=1}^m p_k^0 \prod_{i=1}^{n-1} f_k(x_i)},$$

$j = 1, \ldots, m.$

Second, view these probabilities as prior ones and recompute the posterior as

$$\hat{p}_j^2 = \frac{\hat{p}_j^1 f_j(x_n)}{\sum_{k=1}^m \hat{p}_k^1 f_k(x_m)} .$$

Show that for all $j = 1, \ldots, m$, $\hat{p}_j^2 = p_j^1$.

7. ***Optimal inspection–repair policy of a partially renewed system***
A system has three states: 0, 1 and 2. State 0 corresponds to a new system, state 1 is a "dangerous" state, and state 2 is the failure state. System evolution in time is described by a continuous-time Markov process $\xi(t)$ with time-dependent transition rates: $\lambda_{01}(t) = t$, $\lambda_{12}(t) = t$. All other transition rates equal zero. The system starts operating from state 0. The next inspection is planned after time T. If the inspection reveals state 1, $\xi(t)$ is shifted into state 0, and a new inspection is planned after time T. If the inspection reveals failure, the process stops.

The following costs are introduced: the inspection cost is $c_{ins} = 1$, the repair (shift) cost is $c_{rep} = 2$. Each unit of time the system is in the failure state costs $B = 5$, and the cost of failure is $A_f = 10$.
a. Derive the transition probabilities for $\xi(t)$.
b. Derive the recurrence relation for the minimum costs to find the optimal inspection period T^*.
Hint. Follow the reasoning of Section 5.4.4 to derive the following system of differential equations for $P_{0i}(t) = P(\xi(t) = i | \xi(0) = 0)$:

$$P_{00}'(t) = -\lambda_{0,1}(t) P_{0,0}(t);$$
$$P_{01}'(t) = \lambda_{0,1}(t) P_{0,0}(t) - \lambda_{12}(t) P_{0,1}(t).$$

Solve this system for the initial condition $P_{00}(0) = 1$, $P_{01}(0) = 0$. Note that $P_{02}(t) = 1 - P_{00}(t) - P_{01}(t)$.

8. ***Two-state continuous-time Markov process***
A system has two states: 0 and 1. The transition rate from 0 to 1 is λ, the transition rate from 1 to 0 is μ. The system is in state 0 at time $t = 0$. Denote by $P_0(t)$, $P_1(t)$ the probabilities that at time t the system is in state 0 or 1, respectively. Use the Laplace transform technique to find the probabilities $P_0(t)$, $P_1(t)$. Investigate their behavior for $t \to \infty$.

Solutions to Exercises

The real danger is not that machines will begin to think like men, but that men will begin to think like machines.

Sydney J. Harris

Chapter 1

1a. There are three minimal paths: $\{1,2\}$, $\{1,3\}$, $\{4\}$. By (1.1.4), the structure function is: $\phi(\mathbf{x}) = 1-(1-x_1x_2)(1-x_1x_3)(1-x_4)$.

1b. Multiply the expressions in the first two brackets and use the fact that $x_1^2 = x_1$. Then $\phi(\mathbf{X}) = 1-(1-X_1X_2-X_1X_3+X_1X_2X_3)(1-X_4)$. Now

$r_0 = E[\phi(\mathbf{X})] = 1-(1-p_1p_2-p_1p_3+p_1p_2p_3)(1-p_4) = \psi(p_1,p_2,p_3,p_4)$.

1c. The upper bound is the reliability of a parallel system with independent paths $\{1,2\}$, $\{1^*,3\}$, $\{4\}$. By (1.3.10),

$\phi_{UB}(\mathbf{p}) = 1-(1-p_1p_2)(1-p_1p_3)(1-p_4)$.

Similarly, the lower bound is the reliability of a series connection of two parallel systems which correspond to the minimal cuts $\{1,4\}$, $\{2,3,4^*\}$. Thus

$\phi_{LB}(\mathbf{p}) = (1-(1-p_1)(1-p_4))(1-(1-p_2)(1-p_3)(1-p_4))$.

1d. Substitute $p_i = e^{-\lambda_i t}$ into the expression of r_0. Denote the result by $r_0(t)$. This will be the system reliability. The lifetime distribution function is $F(t) = 1-r_0(t)$.

1e. The stationary availability of element i is $A_i = \mu_i/(\mu_i+\nu_i)$. According to the data, $A_1 = 0.815$, $A_2 = 0.8$, $A_3 = 0.788$, $A_4 = 0.778$. By (1.2.13) and (1.2.16) , system availability is obtained by substituting A_i instead of p_i into the expression for $\psi(p_1,\ldots,p_4)$. The result is $A_v = 0.951$.

2. Differentiate $\psi(\cdot)$ found in Exercise 1b with respect to p_1, p_2 and p_3. The results are:

$R_1 = (1-p_4)(p_2+p_3-p_2p_3)$; $R_2 = (1-p_4)p_1(1-p_3)$,

$R_3 = (1-p_4)p_1(1-p_2)$; $R_4 = 1 - p_1p_2 - p_1p_3 + p_1p_2p_3$.
Substitute the p_i values and obtain that $R_1 = 0.45$, $R_2 = 0.225$, $R_3 = 0.09$, $R_4 = 0.19$. The element 1, therefore, has the highest importance.

3. Consider any vector $\mathbf{x}$. Its ith coordinate is either 1 or 0. In the first case, the left-hand side is $\phi(1_i, \mathbf{x})$, which is equal to the right-hand side. In the second case, the left-hand side is $\phi(0_i, \mathbf{x})$, again equal to the right-hand side.

4. The mean lifetime of the whole system is $(1/n + 1/(n-1) + \ldots + 1)$. The easiest way to obtain this result is the following. The first failure takes place after time $\tau_{\min} = \min(\tau_1, \ldots, \tau_n)$, where $\tau_i \sim \text{Exp}(\lambda = 1)$. It follows that $\tau_{\min} \sim Exp(n)$, and $E[\tau_{\min}] = 1/n$. After the first failure, the situation repeats itself, with n being replaced by $n-1$. Check $n = 3$. The mean lifetime of the parallel system will be $1/3 + 1/2 + 1 = 11/6$. Already for $n = 4$ we have $1/4 + 1/3 + 1/2 + 1 = 25/12$. The answer: $n \geq 4$.

5. Follows directly from (1.3.4).

6. From the expression (1.2.11), the exact reliability equals $r_0 = 2p^3 + 2p^2 - 5p^4 + 2p^5$. The upper bound is $r^* = 1 - (1-p^2)^2(1-p^3)^2$ because there are two minimal paths with two elements and two minimal paths with three elements. The lower bound is $r_* = [1-(1-p)^2]^2[1-(1-p)^3]^2$ since there are two minimal cuts of size 2 and two minimal cuts of size 3. The numerical results are:

p	r_*	r_0	r^*
0.90	0.978 141	0.978 480	0.997 349
0.95	0.994 758	0.994 781	0.999 807
0.99	0.999 798	0.999 798	1.000 000

7 . For a series system, $\partial r_0/\partial p_i = \prod_{j=1}^{n} p_j/p_i$; see (1.2.4). Thus the most important component has the *smallest* p_i, i.e. it is the first one. Similarly, for a parallel system (see (1.2.5)), the most important component has the largest p_i, i.e. it is the nth component.

9a, b. The radar system is a so-called 2-out-of-3 system. It is operational if either two stations are up and one is down, or all three stations are up. Therefore, its reliability is $R(t) = 3(1-G(t))^2G(t) + (1-G(t))^3 = 3\exp[-2\lambda t](1 - \exp[-\lambda t] + \exp[-3\lambda t])$. After simple algebra, $R(t) = 3e^{-2\lambda t} - 2e^{-3\lambda t}$.

To obtain the mean up time, use (1.3.12). This gives $E[\tau] = 5/(6\lambda)$. By (1.2.13), the stationary availability is $A_v = E[\tau]/(E[\tau] + t_{rep}) = 0.893$.

10. Obviously, $\phi(0,0) = 0$, $\phi(1,0) = \phi(0,1) = 1$. Due to the error in designing the power supply system, $\phi(1,1) = 0$, and thus this system is ***not*** monotone; compare with Definition 1.1.1. Note e.g. that here {1} is a minimal path set, but {1,2} is ***not*** a path set.

Chapter 2

1. $\tau_s = \tau_1 + \ldots + \tau_5$, where τ_i are the lifetimes of the units. The result follows from Corollary 2.1.1.

2. By (2.2.3), $h(t) = 1/(1-t)$ for $t \in [0,1)$, since the density of τ is $f(t) = 1$, and $F(t) = t$.

3. The c.d.f. of the system is $F_s(t) = (1 - e^{-5t})(1 - e^{-t})$; see (1.3.5). Let us investigate the failure rate using *Mathematica*.

In the following printout "F" is the c.d.f., "f" the density and "h" the failure rate. Obviously, $h(t)$ is not a monotone function. It is easy to establish that $\lim_{t\to 0} h(t) = 0$ and that $\lim_{t\to\infty} h(t) = 1$.

```
In[4]:= F = (1 - Exp[-5 t]) (1 - Exp[-t]); f = D[F, t];
        h = f / (1 - F);
        Plot[h, {t, 0, 2}, PlotStyle → {AbsoluteThickness[1.2]},
         AxesLabel → {"t", "h(t)"}]
```

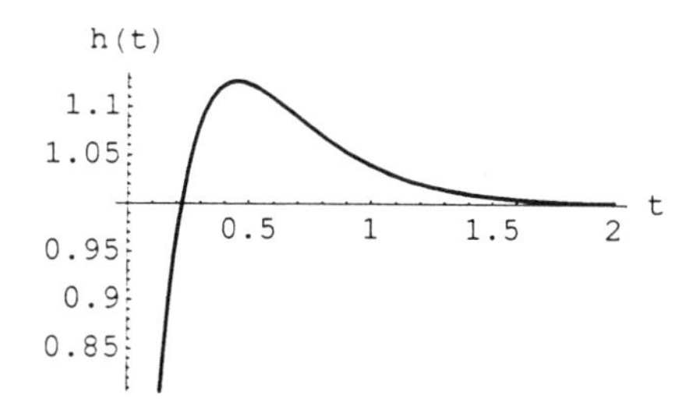


```
Out[6]= - Graphics -
```

5. Here it is easy to obtain an analytic solution. $f(t) = e^{-5t} + 0.8e^{-t}$. After simple algebra, $h(t) = 1+0.8/(0.2+0.8e^{4t})$. Clearly, $h(t)$ *decreases* as t increases.

6a. The coefficient of variation is $c.v. = \sqrt{Var[\tau]}/E[\tau] = 0.4$. Since the solution involves the gamma function (see (2.3.14)), let us use *Mathematica*. The computation details are given in the following printout. The results are

$\lambda = 0.889$ and $\beta = 2.696$.

```
In[1]:= lhs = (Gamma[1 + 2 / β] / (Gamma[1 + 1 / β]) ^2 - 1) ^0.5 - 0.4;
        FindRoot[lhs == 0, {β, 2.5}]

Out[2]= {β → 2.69562}

In[3]:= λ = Gamma[1 + 1 / 2.69562]

Out[3]= 0.889234
```

6b. By (2.3.7), $\sqrt{e^{\sigma^2} - 1} = 0.4$, whence $\sigma = 0.385$. To find μ, we have to solve the equation $1 = \exp(\mu + 0.385^2/2)$. The solution is $\mu = -0.0741$.

7. The group of two elements in parallel has c.d.f. $F_{b,c}(t) = (1 - e^{-2t})^2$; see (1.3.5). The failure rate of this group is $h_{b,c}(t) = 4(1 - e^{-2t})e^{-2t}/(1 - F_{b,c}(t))$.

The failure rate of element a is $h_a(t) = 2t$; see (2.3.11). The whole system has failure rate $h_s(t) = h_a(t) + h_{b,c}(t)$.

8. By (2.3.12), the mean lifetime of a is $\mu_a = \Gamma(1.5) = 0.886$. So, its reliability is at least as great as the reliability of an exponential element with $\lambda_a = 1/0.886$. By Corollary 2.2.6, the lower bound on system reliability is $\psi_L(p_a, p_b, p_c) = p_a(1 - (1 - p_b)^2)$, where $p_a = \exp(-t/\mu_a)$ and $p_b = \exp(-2t)$. This bound is valid for $t \in [0, \min(0.886, 0.5)] = [0, 0.5]$.

9. Substitute $h(u)$ from (2.2.3) into (2.2.5). Represent $f(u)du/(1 - F(u))$ as $-d\log(1 - F(u))$. Then integrate.

10. Consider the Laplace transform of S:

$$L(S) = E[e^{-zS}] = \sum_{k=1}^{\infty} P(K = k)E[e^{-z(Y_1+\ldots+Y_K)}|K = k]$$

$$= \sum_{k=1}^{\infty}(1-p)^{k-1}pE[e^{-z(Y_1+\ldots+Y_k)}] = \sum_{k=1}^{\infty}(1-p)^{k-1}p(E[e^{-zY_1}])^k$$

$$= \sum_{k=1}^{\infty}(1-p)^{k-1}p(\lambda/(\lambda + z))^k p\lambda/(z + p\lambda).$$

After further algebra, we obtain $L(S) = p\lambda/(z + p\lambda)$, which means that $S \sim Exp(p/\mu)$, where $1/\mu = \lambda$.

11. $P(X \le t) = P(\log \tau \le t) = P(\tau \le e^t) = 1 - \exp[-(\lambda e^t)^\beta]$.
Now substitute $\lambda = e^{-a}$ and $\beta = 1/b$: $P(X \le T) = 1 - \exp[-e^{(t-a)/b}]$.

12. By (1.3.4), $P(\tau \le t) = 1 - \prod_{i=1}^{n} e^{-(\lambda_i t)^\beta} = 1 - \exp[-(\lambda_0 t)^\beta]$, where $\lambda_0 = (\sum_{i=1}^{n} \lambda_i^\beta)^{1/\beta}$.

13. A bridge has two minimal cuts of minimal size 2, $\{1,2\}$ and $\{4,5\}$. This explains the expression $g(\theta)$.

15a. For a series system, $F(t) = 1 - \exp[-\lambda_1 t - (0.906t)^4]$. The mean of τ_2 is $\Gamma[1 + 1/4]/0.906 = 1$. The system lifetime density function is $f(t) = \exp[-\lambda_1 t - (0.906t)^4](\lambda_1 + 0.906^4 \cdot 4t^3)$. It is difficult to analyze the behavior of $f(t)$ analytically. The graphical investigation provides interesting results, as is shown below.
15b. The printout shows the graphs of $f(t)$ together with the exponential density of τ_1 (thick curves), for various λ_1 values. It is seen that for $\lambda_1 \ge 2$, the graph of $f(t)$ almost coincides with the exponential density.

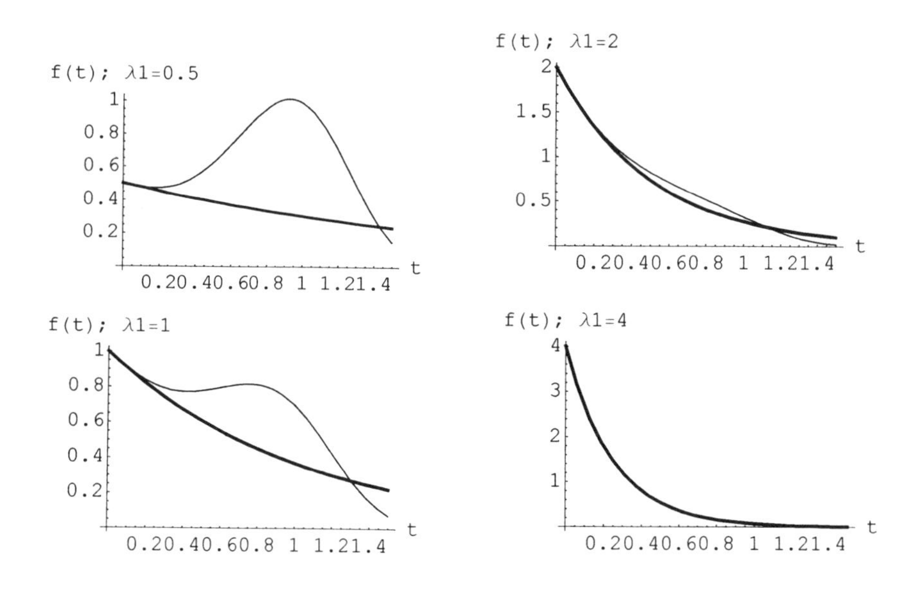

When λ_1 is large, i.e. when the mean value of the exponential component is small, the exponent becomes the dominant component of the density function, because the Weibull component has density close to zero for small values of t. On the other hand, for small λ_1 values, the Weibull component prevails and $f(t)$ has a clearly expressed mode near its mean value.

Chapter 3

1. According to (3.2.2), the 0.1 quantile of τ_1 is the root of the equation $1 - e^{-(\lambda t_{0.1})^\beta} = 0.1$. After some algebra, $t_{0.1} = (\lambda)^{-1}(-\log 0.9)^{1/\beta}$. Substitute the values of λ and β from the solution of Exercise 2.6a. The result is $t_{0.1} = 0.488$.

Denote by $q_{0.1}$ the 0.1 quantile of τ_2. This is the root of the equation $\Phi((\log q_{0.1} - \mu)/\sigma) = 0.1$. Hence $(\log q_{0.1} - \mu)/\sigma = \Phi^{-1}(0.1) = -1.2816$. Using the result of Exercise 2.6b, $q_{0.1} = \exp(\mu - 1.2816\sigma) = 0.567$.

2c. Let us carry out a straightforward maximization of the likelihood function, using *Mathematica*. Let $t_1, \ldots, t_{10}$ be the observed lifetime values in an increasing order. There are three observations censored at $t_{10} = 3$. The likelihood function for the Weibull case is, according to (3.3.9),

$Lik = \prod_{i=1}^{10} (\lambda^\beta \beta t_i^{\beta-1} \exp[-(\lambda t_i)^\beta]) \exp[-3(\lambda t_{10})^\beta]$.

Simplify this expression by taking its logarithm. The contour plot of "logLik" shows a clear maximum point in the neighborhood of $\lambda = 0.4$, $\beta = 1.4$. The operator "FindMinimum" applied to the negative of "logLik" gives $\hat{\lambda} = 0.44$ and $\hat{\beta} = 1.42$.

```
ContourPlot[loglik, {λ, 0.35, 0.65}, {β, 1.2, 1.6}]
```

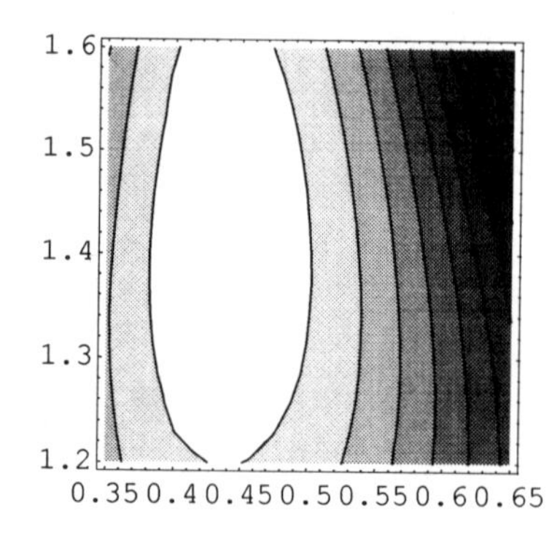


```
Out[4]= - ContourGraphics -

In[5]:= FindMinimum[-loglik, {λ, 0.4}, {β, 1.4}]

Out[5]= {17.6335, {λ → 0.439918, β → 1.41745}}
```

3. Here we have quantal-type data. From (3.3.12),

$Lik = (1 - e^{-(2\lambda)^\beta})^4 (e^{-(2\lambda)^\beta})^6$.

Substitute $\lambda = 0.4$. Then $logLik = 4\log(1 - e^{-(0.8)^\beta}) - 6(0.8)^\beta$. The equation $\partial logLik/\partial\beta = 0$ leads, after some algebra, to the equation $e^{0.8^\beta} - 1 = 1/1.5$. Its solution is $\hat{\beta} = 3.01$.

4. The likelihood is

$$Lik = \lambda^n \exp[-\lambda \sum_{i=1}^{n} (t_i - a)].$$

In this expression, $a \le t_{(1)} = \min(t_1, t_2, \ldots, t_n)$. Thus the largest possible value for a is $t_{(1)}$, the smallest observed lifetime. It is seen from the expression for Lik that it is maximized by $\hat{a} = t_{(1)}$. Proceed in the usual way and derive that the MLE of λ is $\hat{\lambda} = n/\sum_{i=1}^{n}(t_i - \hat{a})$.

5. The likelihood function is $Lik = [F(T;\alpha,\beta)]^{k_1}[F(2T;\alpha,\beta) - F(T;\alpha,\beta)]^{k_2} \times [1 - F(2T;\alpha,\beta)]^{(n-k_1-k_2)}$.

6b. The likelihood function is

$$Lik = \beta^5 \prod_{i=1}^{5} d_i^{\beta-1} \exp[-\sum_{i=1}^{6} d_i^{\beta}],$$

where $d_6 = T - t_6$. The printout shows the graph of the logarithm of the likelihood (denoted as "logLik") with a maximum near 1.4; see *Out[2]*. The "FindMinimum" operator applied to "$-$logLik" gives the MLE $\hat{\beta} = 1.312$; see *Out[3]*.

```
In[1]:=
        logLik = 5 Log[β] + (β - 1) Log[2 * 1.3 * 2.3 * 1.6 * 1.7 * 1.1] -
           2^β - 1.3^β - 2.3^β - 1.6^β - 1.7^β - 1.1^β;
        Plot[logLik, {β, 0.5, 3}]
```

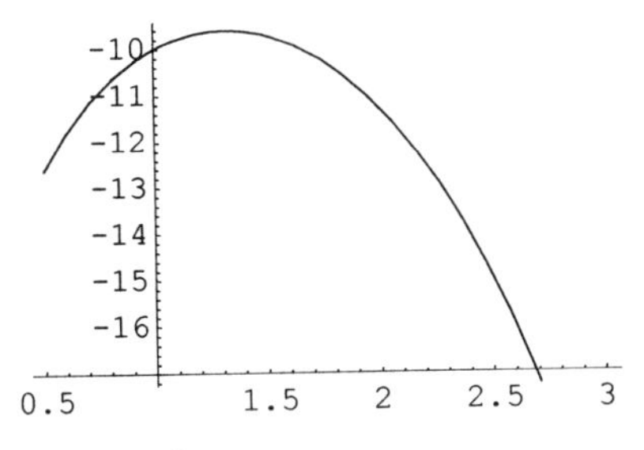


```
Out[2]= - Graphics -

In[3]:= FindMinimum[-logLik, {β, 1.4}]

Out[3]= {9.61019, {β → 1.3226}}
```

7b. The hazard plot is presented in the printout below. It seems plausible that the true hazard plot is a convex function.

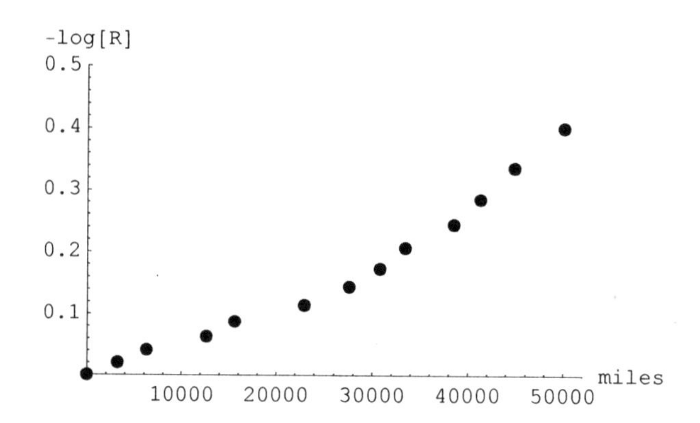

Chapter 4

1a. We have two equations:

$\Phi((\log 2-\mu)/\sigma)=8/127=0.063$ and $\Phi((\log 2-\mu)/\sigma)=58/127=0.457$. Now $\Phi^{-1}(0.063)=-1.53$, $\Phi^{-1}(0.457)=-0.108$. So we have $\log 2=\mu-1.53\sigma$, $\log 5=\mu-0.108\sigma$. Solving these gives $\hat{\mu}=1.678$, $\hat{\sigma}=0.644$. By (2.3.5), the mean of the lognormal distribution is $\exp[\hat{\mu}+\hat{\sigma}^2/2]=6.59$.

1b. By (4.2.11), the cost of age replacement at age $T=3$ is

$$\eta_{age}(3)=\frac{3500F(3)+1000(1-F(3))}{\int_0^3(1-F(t))dt}.$$

The denominator of this expression can also be represented in the following form (integrate by parts):

$$(1-F(3))3+\int_0^3 tf(t)dt,$$

where $f(t)=(\sqrt{2\pi}\sigma t)^{-1}\exp[-(\log t-\mu)^2/2\sigma^2]$ is the lognormal density. $F(3)=\Phi((\log 3-\hat{\mu})/\hat{\sigma})=0.184$. The result of the integration is 0.403 896.
Now $\eta_{age}(3)=(0.184\times3500+0.816\times1000)/(3\times0.816+0.4039)=512$. Suppose that we replace the compressor only at failure. Then we pay, on average, \$3500 per 6.59 years, or \$531 yearly. This is only 4% higher than the cost of age replacement at $T=3$.

2. The mean of $Gamma(k=7,\lambda=1)$ is $\mu_1=k/\lambda=7$; see (2.1.17). The mean of $Gamma(k=4,\lambda=0.5)$ is $\mu_2=8$. Thus, strategy (i) costs $2\times1500/\mu_1=\$429$ and $2\times1500/\mu_2=\$375$ for the first and the second c.d.f., respectively.

Strategy (ii) means that we replace both parts, on average, once per interval of length $\mu_{\min}=E[\min(\tau_1,\tau_2)]$, where $\tau_i\sim Gamma(7,1)$ or $\sim Gamma(4,0.5)$.

For the first distribution, numerical integration shows that $\mu_{\min} = 5.53$. For the second distribution, $\mu_{\min} = 5.81$. Thus strategy (ii) costs 2000/5.53 =\$ 362 and 2000/5.81 =\$344, for the first and the second c.d.f., respectively. Thus strategy (ii) always costs less than strategy (i).

3. The formula for the average costs of age replacement is (4.2.11). We have to compare $c_e = 20$ with $c_e = 5$. The printout below shows two graphs: $\eta_{age}(T)$ for $c_e = 20$, the upper curve, and for $c_e = 5$, the lower curve. The choice is between the optimal age for the upper curve, $T_1^* = 0.36$, and $T_2^* = 0.54$ for the lower curve. The worst-case cost for T_1^* is about 3.67, while the worst-case cost for T_2^* is about 4.80. I would prefer $T_1^* = 0.36$.

```
In[4]:= Num1 = (1 - Exp[-T2^4]) * 5 + Exp[-T2^4];
        Num2 = (1 - Exp[-T2^4]) * 20 + Exp[-T2^4];
        Den = Integrate[Exp[-t^4], {t, 0.0, T2}];
        η1 = Num1 / Den;
        η2 = Num2 / Den;
        Plot[{η1, η2}, {T2, 0.1, 1.5}, PlotRange -> {{0, 1}, {0, 8}},
         GridLines → {{0.25, 0.5, 0.75, 1.0}, {2, 4, 6, 8, 10}},
         AxesLabel → {"T", "η1(T),η2(T)"}]
```

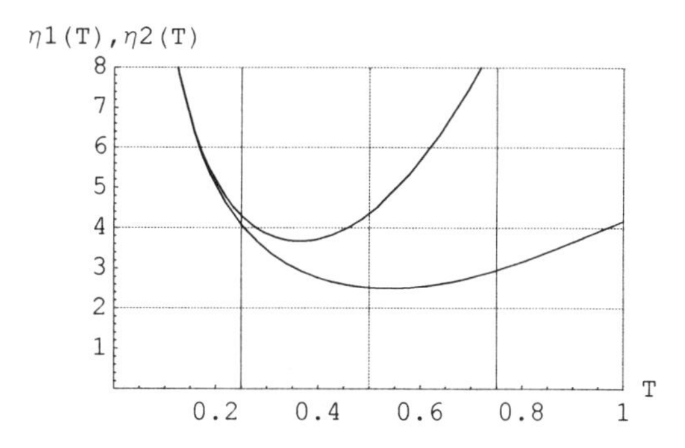


```
Out[9]= - Graphics -

In[10]:= FindMinimum[η1, {T2, 0.4}]

Out[10]= {2.49712, {T2 → 0.538402}}

In[11]:= FindMinimum[η2, {T2, 0.4}]

Out[11]= {3.66842, {T2 → 0.364101}}
```

4. We have a block replacement with random period T. The corresponding expression for the cost criterion is

$$\eta = \frac{\int_a^b f(T)(100m(T)+5)dT}{E[T]} .$$

Since $T \sim N(\mu = 1, \sigma = 0.1)$, we may take $a = \mu - 4\sigma = 0.6$ and $b = \mu + 4\sigma = 1.4$. The denominator equals $\mu = 1$. Let us use the approximation to $m(T)$ based on the formula given in Example 4.1.2:

$m(T) = \sum_{n=1}^{m} Gamma(T; nk, \lambda)$.

It is enough for our case to take $m = 4$. The computation results are presented in the printout below.

"lb" is the approximation to the renewal function from below, "ub" from above. The result is $\eta = 6.960\,89$.

The graph shows the plots of $\eta_b(T)$ with $m(T)$ replaced by the upper and lower bound, respectively ("UB" and "LB"). In the neighborhood of the minimum point (near $T^* = 1$), both curves are seen to coincide.

It is worth noting that in the absence of periodic replacement, the cost will be $\$100/E[\tau] = 25$, since the mean lifetime of the street lamp is $E[\tau] = k/\lambda = 4$.

```
In[1]:= g4 = 1 - Exp[-t] * Sum[t^i / i!, {i, 0, 3}];
        g8 = 1 - Exp[-t] * Sum[t^i / i!, {i, 0, 7}];
        g12 = 1 - Exp[-t] * Sum[t^i / i!, {i, 0, 11}];
        g16 = 1 - Exp[-t] * Sum[t^i / i!, {i, 0, 15}];
        lb = g4 + g8 + g12 + g16;
        ub = lb + g4^5 / (1 - g4);
        η = 100 * (Sqrt[2 Pi] * 0.1)^(-1) *
           NIntegrate[lb * Exp[-(t - 1)^2 / 0.02], {t, 0.6, 1.4}] + 5

Out[7]= 6.96089

In[8]:= Plot[{(100 * lb + 5) / t, (100 * ub + 5) / t}, {t, 0.1, 6},
         GridLines → Automatic, AxesLabel → {"t", "UB, LB"}]
```

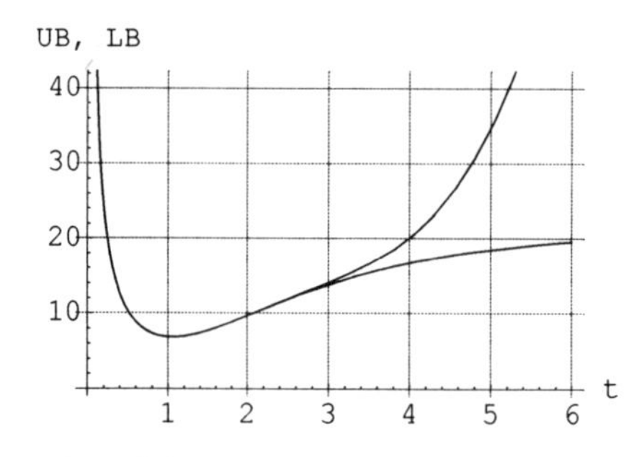


```
Out[8]= - Graphics -
```

5. First, find β and λ. From (2.3.14) it follows that $\beta = 3.714$. From (2.3.12) we obtain $\lambda = 0.000451$. The formula for the stationary availability is

$$A_v(T) = \frac{\int_0^T (1 - F(t;\lambda,\beta))dt}{\int_0^T (1 - F(t;\lambda,\beta))dt + 50F(T;\lambda,\beta) + 10(1 - F(T;\lambda,\beta))} .$$

After simple algebra,

$$A_v(T) = 1/\Big(1 + \frac{50F(T;\lambda,\beta) + 10(1 - F(T;\lambda,\beta))}{\int_0^T (1 - F(t;\lambda;\beta))dt}\Big) .$$

The printout shows the investigation of the fraction in the last expression, denoted as "Num/Den". *Mathematica* does not like using the upper integration limit T as a variable in the Plot and FindMinimum operators, but nevertheless carries out the calculations. The optimal replacement age is $T^* = 1170$ hrs, and the availability is $A_v(T^*) = 1/(1 + 0.0118) = 0.988$.

```
In[18]:= λ = 0.000451; β = 3.714;
         Num = 50 (1 - Exp[- (λ * T) ^β]) + 10 * Exp[- (λ * T) ^β];
         Den = NIntegrate[Exp[- (λ * t) ^β], {t, 0, T}];
         Plot[Num / Den, {T, 100, 4000}, AxesLabel → {"T", "Num/Den"}]

    NIntegrate::nlim : t = T is not a valid limit of integration.
```

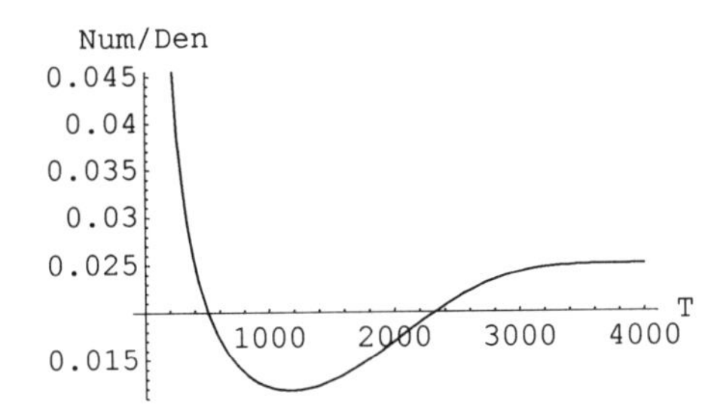


```
Out[21]= - Graphics -

In[22]:= FindMinimum[Num / Den, {T, 100}]

Out[22]= {0.0118129, {T → 1169.79}}
```

6. Step 1: *Optimization at the element (part) level.*
Part 1: The expression for the costs is

$$A_{11}(T) = \frac{10 \int_0^T 0.1tdt + 2}{T} = (0.5T^2 + 2)/T .$$

It is easy to check that the minimum on the grid $T = 1, 2, 4, 8, 16$ is attained at $T = 2$.

Similarly, for part 2,

$$A_{12}(T) = \left(10 \int_0^T 0.06t^2 dt + 2\right)/T = (0.2T^3 + 2)/T,$$

and the optimal $T = 2$.

Step 2: ***Optimization at the system level.***
The basic sequence is $(2,2)$, $(4,4)$, $(8,8)$, $(16,16)$. The expression for the costs is

$$\eta(T) = (0.5T^2 + 2)/T + (0.2T^3 + 2)/T + 8/T.$$

The last term $8/T$ reflects the system set-up costs paid at each planned replacement. It is easy to check that the minimum of $\eta(T)$ is attained at $T = 2$, i.e. for the first vector of the basic sequence. The conclusion: replace both parts every two time units. The optimal value of $\eta(T)$ is 7.8.

7a. The first step is to solve equation (A.2.9) in Appendix A. T_{tot}= T1 +T2, where T1=26 +32.5+ 43.3=101.8 in thousands of miles. Similarly, T2=19.5 + 35.3 + 50.2 + 68.7 = 173.7. n1=3, n2=4, are the number of observed engine failures. The expression "eq" (see the printout) is the left-hand side of (A.2.9) (the fraction is transferred to the left). The operator "FindRoot" produces $\hat{\beta} = 0.024\,044\,7$. Substituting it into (A.2.11) gives $\hat{\alpha} = -3.683\,76$.

```
In[1]:= T = 101.8 + 173.7; n1 = 3; n2 = 4; n = n1 + n2;
        A = Exp[β 50] * 50 + Exp[β 70] * 70;
        B = Exp[β 50] + Exp[β 70];
        eq = T + n / β - n * A / (B - 2);
        FindRoot[eq == 0, {β, 0.05}]

Out[5]= {β → 0.0240447}

In[6]:= β = 0.024; α = Log[β (n1 + n2) / (Exp[β 50] + Exp[β 70] - 2)]

Out[6]= -3.68376
```

7b. The relevant expression for the cost criterion is (4.2.7). Its numerator equals num $= 1000 \int_0^T \exp[\hat{\alpha} + \hat{\beta}t]dt + c_{rep} = 1000e^{\hat{\alpha}}(e^{\hat{\beta}T} - 1)/\hat{\beta} + c_{rep}$, where c_{rep} lies between 100 and 300. The denominator equals T. The expression (4.2.7) for $\eta_D(T)$ is denoted in the printout by η. The plot shows two curves for the costs as a function of T. The upper curve corresponds to $c_{rep} = 300$, the lower – to $c_{rep} = 100$. The optimal repair periods are 16 000 and 25 600, for $c_{rep} = 100$ and $c_{rep} = 300$, respectively. If we assume that $c_{rep} = 100$, and use a near-optimal period of 15 000 miles, the worst-case situation would be \$50.2 per thousand miles. If we assume that $c_{rep} = 300$ and do a minimal repair every $T^* = 25\,000$ miles, the worst-case situation would be \$46.4 per thousand miles. The "minimax" reasoning makes the second option, $T^* = 25\,000$ miles, preferable.

```
In[7]:= α = -3.68376; num = 1000 * Exp[α] * (Exp[β T1] - 1) / β + 100;
        den = T1; η = num / den; η1 = (num + 200) / den;
        Plot[{η, η1}, {T1, 1, 200}]
```

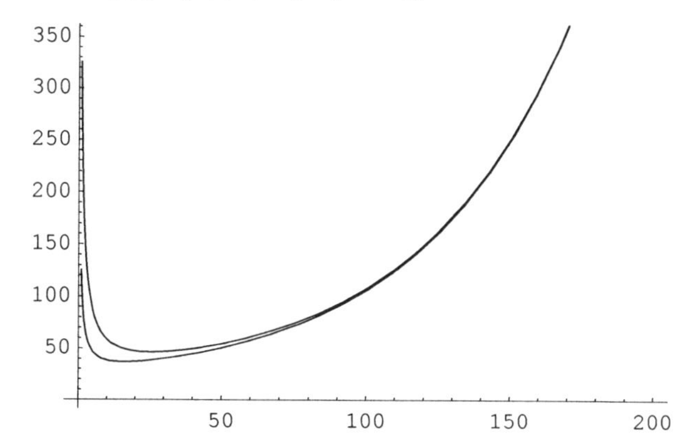


```
Out[9]= - Graphics -

In[10]:= FindMinimum[η, {T1, 100}]

Out[10]= {36.8846, {T1 → 15.9917}}

In[11]:= FindMinimum[η1, {T1, 100}]

Out[11]= {46.4234, {T1 → 25.5754}}

In[12]:= T1 = 15; η // N

Out[12]= 36.9134

In[13]:= η1 // N

Out[13]= 50.2468

In[14]:= T1 = 25; η // N

Out[14]= 38.4308

In[15]:= η1 // N

Out[15]= 46.4308
```

8. The printout below shows the details of the numerical investigation of $\eta(T)$ for $f(x) = e^{-x}$. The c.d.f. $F(x) = 1 - e^{-x}$. The calculations were done for k ranging from 0 to 20. $A(T)$ is denoted as "A" and "A1," and $B(T)$ as "B" and "B1," for $t_0 = 0.05$ and $t_0 = 0.1$, respectively. (t_0 is denoted as "t").

The upper graph shows $\eta(T) = A/B$ for $t_0 = 0.05$. The optimal $T^* \approx 0.27$, and the maximum $\eta(T^*) \approx 0.74$.

The lower graph shows $\eta(T)$ for $t_0 = 0.1$. The optimal $T^* \approx 0.4$, and the maximum $\eta(T^*) \approx 0.65$.

An important observation is that choosing large T values, say $T = 0.8$ (note that this is 80% of the mean CDE lifetime), may considerably reduce the availability.

In my opinion, taking $T = 0.33$, i.e. in the "middle," would provide near-optimal results for any t_0 in the range $[0.05, 0.1]$.

```
In[159]:=
        A = Sum[Integrate[(x - k * t) Exp[-x],
              {x, k * (T + t), k * (T + t) + T}], {k, 0, 20}] +
           Sum[(-Exp[-(k + 1) * (T + t)] + Exp[-(k + 1) (T + t) + t]) *
              (k + 1) T, {k, 0, 20}]; t = 0.05;
        B = Sum[(k + 1) * (T + t) (Exp[-k * (T + t)] -
               Exp[-(k + 1) * (T + t)]),
           {k, 0, 20}]; t = 0.05;

        A1 = Sum[Integrate[(x - k * t) Exp[-x],
              {x, k * (T + t), k * (T + t) + T}], {k, 0, 20}] +
           Sum[(-Exp[-(k + 1) * (T + t)] + Exp[-(k + 1) (T + t) + t]) *
              (k + 1) T, {k, 0, 20}]; t = 0.1;
        B1 = Sum[(k + 1) * (T + t)
             (Exp[-k * (T + t)] - Exp[-(k + 1) * (T + t)]),
           {k, 0, 20}]; t = 0.1;
        Plot[{A / B, A1 / B1}, {T, 0.1, 0.8},
         PlotRange → {{0, 0.8}, {0.5, 0.8}},
         AxesLabel → {"T", "η(T)"}, GridLines →
          {{0.2, 0.4, 0.6, 0.8}, {0.5, 0.6, 0.7, 0.8}}]
```

η(T)
0.8
0.75
0.7
0.65
0.6
0.55
0.10.20.30.40.50.60.70.8 T

```
Out[163]= - Graphics -
```

9. The mixture of exponential lifetimes is a DFR distribution; see Theorem 2.2.3. Therefore, no finite maintenance period can be optimal. The answer is $T^* = \infty$, i.e. never carry out the PM.

10. By (2.3.12), $E[\tau_2] = \Gamma[1 + 1/2] = 0.886$. Thus λ_1 should be $1/0.886 = 1.13$. The lifetime of the series system is $F_s(t) = 1 - e^{-1.13t}e^{-t^2}$.
The expression for $\eta(T)$ is

$\eta(T) = (F_s(T) + (1 - F_s(T))c_p)/ \int_0^T (1 - F_s(t))dt.$

This expression is defined in $In[93]$; see the printout below. The graph of $\eta(T)$ is shown in $Out[95]$. The "FindMinimum" operator $In[96]$ provides $T^* = 0.565$ and the minimal $\eta(T^*) = 1.808$. Recall that the efficiency $Q = 1/(E[\tau_s]\eta(T^*))$. $Out[98]$ shows that $Q = 1.07$, quite a low efficiency.

The graph in $Out[100]$ shows the function $\eta(T)$ for a system without the exponential component. The optimal T is 0.511, and the optimal η is 0.817. The corresponding efficiency (denoted as Q) is now 1.38, which is considerably higher than the previous value of 1.07.

How can this fact be explained? The presence of an exponential component changes the form of the density function near zero and makes it very close to the exponential density. We advise the reader to reexamine Exercise 15 in Chap. 2. The presence of an exponential component with mean life equal to or less than the mean life of the aging Weibull component would make Q near 1. We call this phenomenon "exponential minimum-type contamination." It reduces drastically the efficiency of the optimal age replacement.

The practical conclusion is the following. If there is a suspicion that the series system contains many exponentially distributed components, it is not advisable to replace preventively the *whole* system. The PM should be applied only to that part of the system which has an increasing failure rate. In our example, this is the second component whose life follows the Weibull distribution with $\beta = 2$.

```
In[93]:= c = 0.2; λ = 1.13; F1 = 1 - Exp[-λ * y];
         F2 = 1 - Exp[-y^2]; Fs = 1 - (1 - F1) * (1 - F2);
         num = 1 - Exp[-λ * T - T^2] + c * Exp[-λ * T - T^2];
         den = NIntegrate[1 - Fs, {y, 0, T}]; η = num / den;
         Plot[η, {T, 0.1, 5}, AxesLabel → {"T", "η(T)"}]
```

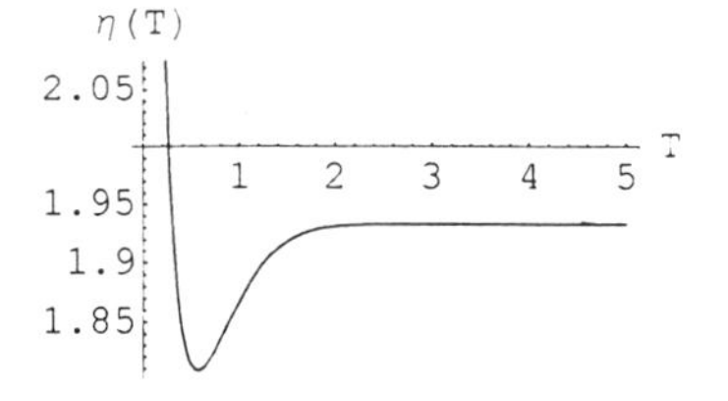


```
Out[95]= - Graphics -
```

```
In[96]:= FindMinimum[η, {T, 0.5}]

Out[96]= {1.80829, {T → 0.565179}}

In[98]:= Q = 1 / (NIntegrate[1 - Fs, {y, 0, 5}] * 1.808)

Out[98]= 1.06899

In[99]:= c = 0.2; λ = 1.13; F2 = 1 - Exp[-y^2];
         num = 1 - Exp[-T^2] + c * Exp[-T^2];
         den = NIntegrate[Exp[-y^2], {y, 0, T}]; η = num / den;
         Plot[η, {T, 0.1, 5}, AxesLabel → {"T", "η(T)"}]
```



```
Out[100]= - Graphics -

In[101]:= FindMinimum[η, {T, 0.5}]

Out[101]= {0.817048, {T → 0.510655}}

In[102]:= Q = 1 / (NIntegrate[Exp[-y^2], {y, 0, 5}] * 0.817)

Out[102]= 1.38113
```

11. Check that $\sigma = 0.246$, $\mu = -0.030\,258$, and $\lambda = 0.913$, $\beta = 4.5$ give that a c.v. of 0.25 and a mean of 1 for both distributions. Recall that for $\tau \sim logN(\mu, \sigma)$, the c.d.f. is $\Phi((log(t) - \mu)/\sigma)$. The corresponding calculations are carried out in $In[134]$; see the printout below. Note that one has to call the "ContinuousDistributions" package and define afterwards the standard normal c.d.f. as *NormalDistribution[0,1]*. $Out[141]$ presents the graphs of $\eta_{age}(T)$ for the lognormal and the Weibull cases. Both curves practically coincide so that maintenance age $T^* \approx 0.6$ for both cases.

```
In[134]:= << Statistics`ContinuousDistributions`
          ndist = NormalDistribution[0, 1];
          c = 0.2; μ = -0.030258; σ = 0.246;
          num = (1 - c) * CDF[ndist, (Log[T] - μ) / σ] + c;
          den = NIntegrate[
            CDF[ndist, - (Log[x] - μ) / σ], {x, 0.0001, T}];
          λ = 0.913; β = 4.54;
          mum1 = (1 - c) * (1 - Exp[- (λ * T) ^ β]) + c;
          den1 = NIntegrate[Exp[- (λ * z) ^ β], {z, 0, T}];
          Plot[{num / den, mum1 / den1}, {T, 0.1, 1.5}]
```

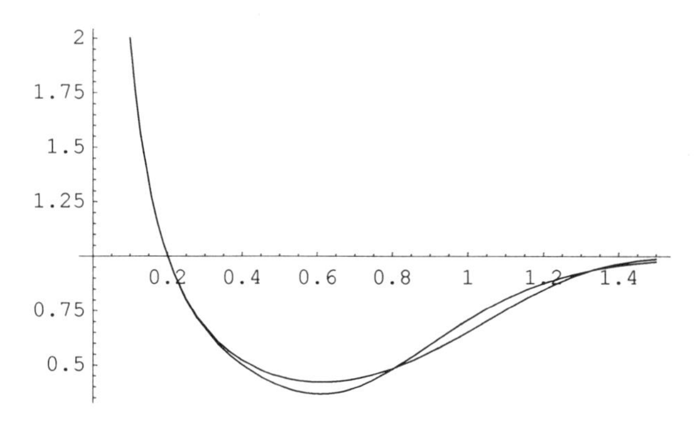


```
Out[141]= - Graphics -
```

12. The "FindMinimum" operator applied to $S = \sum_{i=1}^{5}(\log n_i - a - \alpha(i-1))^2$ gives $\hat{a} = 1.93, \hat{\alpha} = 0.277$. The "Plot" operator produces the graph of $\eta(K)$ shown below. The optimal $K^* = 5$, and the minimal costs are about \$500 per cycle.

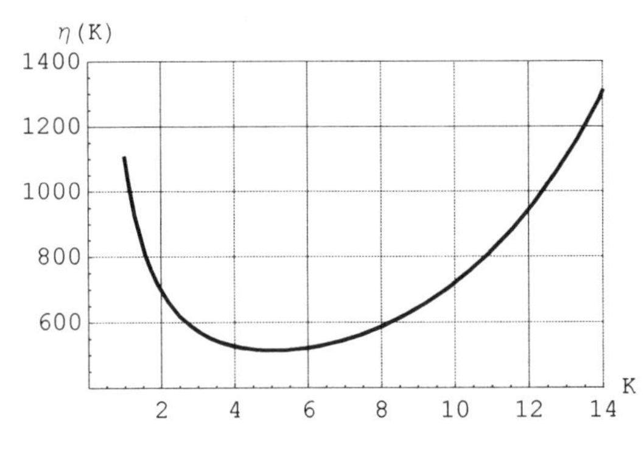


```
- Graphics -
```

Chapters 5–7

1. The system of equations for finding the stationary probabilities is (5.4.5). In our case, the system is

$\pi_1 = 0.1\pi_2$; $\pi_2 = 0.3\pi_1 + \pi_3$; $\pi_1 + \pi_2 + \pi_3 = 1$.

Its solution is presented in the printout below: $\pi_1 = 0.048$, $\pi_2 = 0.483$, $\pi_3 = 0.469$. (Note that one of the equations of the system $\pi = \pi\mathbf{P}$ can be deleted. We deleted the third one.) The mean one-step rewards are equal to 1, and the mean transition times are: $\nu_1 = \nu_2 = 1, \nu_3 = 2$. By (5.4.14), $g = 1/(\pi_1 + \pi_2 + 2\pi_3) = 0.681$.

```
In[165]:= Solve[{π1 - 0.1 * π2 == 0, π2 - 0.3 * π1 - π3 == 0,
             π1  + π2 + π3 - 1 == 0}, {π1, π2, π3}]

Out[165]= {{π1 → 0.0483092, π2 → 0.483092, π3 → 0.468599}}

In[168]:= g = 1 / (0.0483 + 0.483 + 2 * 0.469)

Out[168]= 0.680596
```

2a. The mean length of one operation cycle is $E[\tau_{(k)}] = \mu_{(k)}$. On this cycle, the mean total "up" time of all lines is

$$E[W(k)] = n\mu_{(1)} + (n-1)(\mu_{(2)} - \mu_{(1)}) + \ldots + (n-k+1)(\mu_{(k)} - \mu_{(k-1)}),$$

see (5.2.1). The mean total operation time of all lines during one cycle is $n\mu_{(k)}$ and thus the mean total idle time is $E[\text{Idle}] = n\mu_{(k)} - E[W(k)]$. Thus the mean reward per unit time is

$$\chi_k = \frac{c_{rew} \cdot E[W(k)] - c_f \cdot E[\text{Idle}] - c_{rep}}{\mu_{(k)}}.$$

Let $f_i(t)$ be the density of the ith order statistic:

$$f_i(t) = \frac{n!}{(i-1)!(n-i)!}[F(t)]^{i-1}[1-F(t)]^{n-i} f(t).$$

The derivation of this is instructive. The ith ordered observation lies in the interval $(t, t+dt)$ if (i) a single observation lies in this interval; (ii) any $i-1$ out of the remaining $n-1$ observations are less or equal than t.

The probability of (i) is $\binom{n}{1} f(t)dt$; the probability of (ii) is given by the expression $\binom{n-1}{i-1}[F(t)]^{i-1}[1-F(t)]^{n-i}$. To find $\mu_{(i)}$, we have to integrate

$$\mu_{(i)} = \int_0^\infty t f_i(t)dt \, .$$

Using the *Mathematica* operator " NIntegrate" is very convenient for the numerical evaluation of $\mu_{(i)}$.

2b The printout below details the computation. $F(t) = 1 - \exp[-(0.25t)^{\beta}]$. The density $f(t)$ is found using the operator "D[F,t]." f1, f2, f3 and f4 are defined according to the formula above. $\mu 1$, $\mu 2$, $\mu 3$ and $\mu 4$ are the mean values of the order statistics. The upper limit 40 serves as "infinity." The printout shows the values of $\chi 1, \ldots, \chi 4$. The optimal choice is $k = 2$ which guarantees the highest reward 2.74. The choice $k = 1$ is the second best with reward 2.60. The worst choice is $k = 4$, which would reduce the reward to 1.80.

```
In[183]:= F = 1 - Exp[- (0.25 * v) ^1.5] ; f = D[F, v] ;
          f1 = 4 ! ((1 - F) ^3) * f / 3 ! ;
          f2 = 4 ! * F * ((1 - F) ^2) * f / 2 ! ;
          f3 = 4 ! F^2 * (1 - F) * f / 2 ! ; f4 = 4 ! (F^3) * f / 3 ! ;
          μ1 = NIntegrate[v * f1, {v, 0, 40}]
          μ2 = NIntegrate[v * f2, {v, 0, 40}]
          μ3 = NIntegrate[v * f3; {v, 0, 40}]
          μ4 = NIntegrate[v * f4, {v, 0, 40}]

Out[186]= 1.43302

Out[187]= 2.64486

Out[188]= 4.05988

Out[189]= 6.30617

In[198]:= x1 = (4 * 1.433 - 2) / 1.433
          x2 =
           ((4 * 1.433 + 3 * (2.645 - 1.433)) 1.1 - 0.1 * 4 * 2.644 - 2) /
           2.645
          x3 = ((4 * 1.433 + 3 * (2.645 - 1.433) + 2 * (4.06 - 2.65)) 1.1 -
               0.1 * 4 * 4.06 - 2) / 4.06
          x4 = ((4 * 1.433 + 3 * (2.645 - 1.433) + 2 * (4.06 - 2.65) +
                  (6.31 - 4.06)) 1.1 - 0.1 * 4 * 6.31 - 2) / 6.31

Out[198]= 2.60433

Out[199]= 2.73996

Out[200]= 2.40956

Out[201]= 1.79997
```

3. In order to be able to process the data, we need to call "DescriptiveStatistics." The data are presented in the form of points (hours, flights) in the (H, F) plane.

The operator "For" investigates the coefficient of variation of the lifetime, which is defined as a linear combination of two time scales, hours (H), and flights (F): Tdata=$(1-a)\times$ Hdata + a $\times$ Fdata; $a = 0.0(0.05)1.00$. The output (only the cases $a = 0, a = 0.7$ and $a = 1$ are displayed) shows that $a = 0.7$ provides the smallest c.v. (the c.v. is denoted by "cvT"). It equals 0.082, while for the lifetime on the number-of-flights scale it equals 0.204.

To compute the optimal replacement age, we use the formula (6.4.13). Note that $\hat{F}_a(t)$ is a piecewise constant function. For all three scales $\mathcal{H}, \mathcal{F}, \mathcal{T}_{0.7}$, the optimal replacement age coincides with the smallest observed lifetime.

The results are:

$\hat{\gamma} = 0.302$, for the $\mathcal{H}$ scale; $x^* = 1950$ hours;

$\hat{\gamma} = 0.314$, in the $\mathcal{F}$ scale; $x^* = 920$ flights;

$\hat{\gamma} = 0.229$, in the $\mathcal{T}_{0.7}$ scale, $x^* = 1649$.

We see, therefore, that using the scale with the smallest c.v. provides a considerable reduction in the mean cost γ.

```
In[217]:= HFdata = {{3700, 920}, {3900, 1020},
              {4200, 1200}, {2700, 1370}, {3100, 1540},
              {1950, 1520}, {2100, 1630}, {2300, 1660},
              {2700, 1760}, {2800, 1840}};
          Needs["Graphics`MultipleListPlot`"]
          ListPlot[HFdata, PlotStyle → PointSize[0.025],
           PlotRange → {{0, 4300}, {0, 2100}},
           AxesLabel → {"H", "F"}]
```

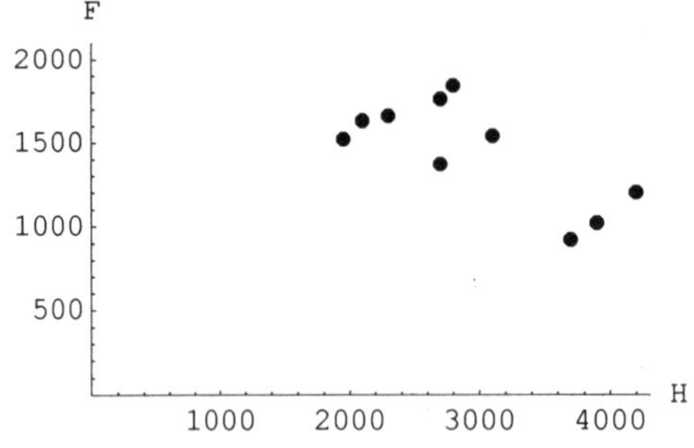


```
Out[219]= - Graphics -

In[220]:= Hdata = Table[{3700, 3900, 4200, 2700,
               3100, 1950, 2100, 2300, 2700, 2800}];
          Fdata = Table[{920, 1020, 1200, 1370,
               1540, 1520, 1630, 1660, 1760, 1840}];
```

```
In[261]:= For[i = 0, i < 21, i++, a = 0.05 * i;
            Tdata = (1 - a) Hdata + a * Fdata;
            cvT = (VarianceMLE[Tdata]) ^0.5 / Mean[Tdata];
            Print["a=", a, "...", "cvT=", cvT]];

      a=0...cvT=0.248186

      a=0.7...cvT=0.0824513

      a=1....cvT=0.204244
```

4. Consider expression (6.3.7). Substitute $x = t_p$ and use the fact that $F(t_p) = p$. Then (6.3.7) takes the form

$$\gamma(t_p) = \frac{p + (1-p)d}{\int_0^{t_p}(1 - F(x))dx}.$$

The function $1 - F(x)$ equals 1 at $x = 0$ and $1 - p$ at $x = t_p$. Assuming linearity, it is easy to obtain that the integral equals $(1 + (1-p))t_p/2$. Substitute this into the previous formula. The result is

$$\gamma(t_p) = \frac{2(p + (1-p)d)}{(2-p)t_p}.$$

Carrying out the PM at age t_p has the advantage that the failure probability before the PM does not exceed a predetermined value p.

5. Suppose we have k nonfailed parts which are put to work for one time period. Denote by $b(k, z)$ the probability that exactly z out of k will fail during this period, $z = 0, 1, \ldots, k$. Obviously,

$$b(k, z) = \binom{k}{z} p^z (1-p)^{k-z}.$$

Suppose that we have ahead of us $m + 1$ time periods and j nonfailed parts. We decide to put to work k_j parts, $k_j \leq j$, after which we follow the optimal policy. The average cost will then be

$$\hat{W}_j(m+1) = b(k_j, k_j) \times 1 + \sum_{z=0}^{k_j - 1} b(k_j, z) W_{j-z}(m).$$

Indeed, we pay 1 for a failure during the unit interval, which happens when all parts put to work have failed during this interval. If z parts fail, $z = 0, 1, \ldots, k_j - 1$, then with probability $b(k_j, z)$ we pay, by the definition of $W_i(m)$, the minimum cost $W_{j-z}(m)$ during the m remaining intervals, since we start with $j - z$ nonfailed parts.

Applying the optimality principle, we write

$$W_j(m+1) = \min_{1 \le k_j \le j} \Big[b(k_j, k_j) + \sum_{z=0}^{k_j-1} b(k_j, z) W_{j-z}(m) \Big].$$

In order to find the optimal policy, we have to apply the "backward motion" algorithm. First, find $W_j(1)$ by assuming that there is only one period ahead, and j nonfailed parts are available, $j = 1, \ldots, r$. (Set $W_j(0) = 0$.) Then, using the values of $W_j(1)$, calculate $W_j(2)$ via the above recurrence relation, etc.

Obviously, $W_j(1) = \min_{1 \le k_j \le j} p^{k_j} = p^j = b(j,j)$. This indicates that $W_j(1)$ is equal to the probability that all j parts will fail on a unit-length interval. Suppose that we are at $t = m - 1$ and there are two unit-length periods ahead. Then

$$W_j(2) = \min_{1 \le k_j \le j} \Big[b(k_j, k_j) + \sum_{z=0}^{k_j-1} b(k_j, z) W_{j-z}(1) \Big].$$

The system failure in $[m - 1, m + 1]$ can occur in two ways: either all k_j parts put to work at $t = m - 1$ fail during $[m - 1, m]$, or z parts fail in the first period, $z < k_j$, and the remaining $j - z$ parts fail during $[m, m + 1]$.

Now it is clear that the expression in the brackets is equal to the probability of system failure during $[m-1, m+1]$, calculated under the following conditions:

(i) The process starts at $t = m - 1$ with j nonfailed parts.

(ii)The first decision is to put k_j parts to work.

(iii) The decision at $t = m$ is made optimally.

This means that $W_j(2)$ is equal to the ***minimum*** probability of a failure in a two-step process. By analogy, it can be shown that $W_j(m)$ gives the desired value of the minimum probability in a m-step process.

6. Substitute $\hat{p}_j^1$, $j = 1, \ldots, m$, into the expression for $\hat{p}_j^2$. The denominator of $\hat{p}_j^1$ cancels out, and we arrive at the formula for p_j^1.

7a. Let the system be in state 0 at $t = 0$. Consider the interval $[t, t + h]$, where $h > 0$, $h \to +0$. The system will be in state 0 at $t + h$ if it was in this state at t and no transition appeared into state 1 during $[t, t + h]$. This leads to the equation:

$$P_{00}(t + h) = P_{00}(t)(1 - \lambda_{01}(t)h) + o(h) .$$

Simple algebra leads to the differential equation

$$P'_{00}(t) = -\lambda_{01}(t) P_{00}(t) .$$

Solving for for $\lambda_{01}(t) = t$ and the initial condition $P_{00}(0) = 1$ gives $P_{00}(t) = \exp[-t^2/2]$. This result was expected: the sitting time in state 0 has the Weibull distribution, since the transition rate to 1 (i.e. the "failure" rate) is $\lambda_{01}(t) = t$.

Relating the probabilities of being in state 1 at time $t+h$ to the probabilities of being either in state 0 or in state 1 at time t, we arrive at the equation

$$P_{01}(t+h) = P_{01}(t)(1-\lambda_{12}(t)h) + P_{00}(t)\lambda_{01}(t)h + o(h).$$

After simple algebra, letting $h \to 0$, we obtain:

$$P'_{01}(t) = -\lambda_{12}(t)P_{01}(t) + \lambda_{01}(t)P_{00}(t).$$

The initial condition is $P_{01}(0) = 0$. Check that the solution is $P_{01}(t) = \exp[-t^2/2]t^2/2$. Thus $P_{02}(t) = 1 - P_{00}(t) - P_{01}(t) = 1 - e^{-t^2/2}(1+t^2/2)$.

Note that $P_{02}(t) = P(\tau_{02} \leq t)$, where τ_{02} is the transition time from state 0 to state 2. The corresponding density function $f_{02}(t)$ is the derivative of $P_{02}(t)$.

7b. Let us derive the recurrence equation for the optimal inspection period.

Denote by W_0 the minimal costs under the optimal inspection policy. Let T be the length of the first inspection period. Then

$$W_0 = \min_{0<T} \Big[c_{ins} + \int_0^T (A + B(T-x)) f_{02}(x)dx$$
$$+ W_0(P_{01}(T) + P_{00}(T)) + c_{rep}P_{01}(T) \Big].$$

Indeed, the inspection cost c_{ins} is always paid. With probability $P_{00}(T)$ the next observed state will be 0 and we will pay in the future only W_0. With probability $P_{01}(T)$ the next observed state is 1, and the cost paid will be $c_{rep} + W_0$, because the process is shifted into state 0. With probability $f_{02}(x)dx$ the process enters the failure state 2 in the interval $(x, x+dx)$ $x \in [0,T]$; the cost associated with this event is $A + B(T-x)$.

Note that the integral in this formula can be simplified and rewritten as $B\int_0^T P_{02}(x)dx + AP_{02}(T)$.

Suppose that the optimal $T = T^*$. Then, by the definition of W_0, it should satisfy the equality:

$$W_0 = \Big(c_{ins} + AP_{02}(T^*) + B\int_0^{T^*} P_{02}(x)dx + c_{rep}P_{01}(T^*) \Big) / P_{02}(T^*).$$

Then, obviously, W_0 and T^* can be obtained by minimizing the right-hand side of the last expression with respect to T. The printout below shows the results of the minimization. The graph suggests that the optimal T is near 1.85. The minimum cost $V_0 = 14.49$.

```
In[7]:= P0 = Exp[-T^2 / 2]; P1 = P0 * T^2 / 2; P2 = 1 - P1 - P0;
        Plot[(1 + 10 * P2 + 5 * NIntegrate[
              1 - Exp[-u^2 / 2] - Exp[-u^2 / 2] * u^2 / 2, {u, 0, T}]) /
          P2, {T, 1, 4}, AxesLabel → {"T", "V"}]
        FindMinimum[(1 + 10 * P2 + 5 * Integrate[1 - Exp[-u^2 / 2] -
               Exp[-u^2 / 2] * u^2 / 2, {u, 0, T}]) / P2, {T, 2}]
```

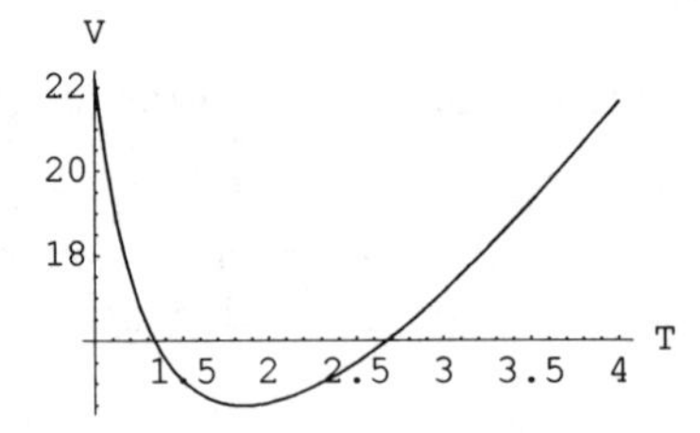


```
Out[8]= - Graphics -

Out[9]= {14.4883, {T → 1.85449}}

        {14.4883, {s → 1.85449}}
```

8. *Two-state continuous-time Markov process.*
The system of differential equations (5.4.21) has, for our problem, the following form:

$$P_0'(t) = -\lambda P_0(t) + \mu P_1(t),$$
$$P_1'(t) = \lambda P_0(t) - \mu P_1(t).$$

Note that $P_0(0) = 1$. We need in fact only the first equation since $P_0(t)+P_1(t) = 1$. Substituting $P_1(t) = 1 - P_0(t)$ into the first equation, we obtain

$$P_0'(t) = \mu - (\lambda + \mu)P_0(t).$$

Applying the Laplace transform to both sides of this equation, we arrive at the equation

$$s\pi_0(s) - 1 = \mu/s - (\lambda + \mu)\pi_0(s),$$

where $\pi_0(s)$ is the Laplace transform of $P_0(s)$ and μ/s is the Laplace transform of μ; see Appendix C. This leads to the formula

$$\pi_0(s) = \frac{s+\mu}{s(s+\lambda+\mu)} = \frac{1}{s+\lambda+\mu} + \frac{\mu}{s(s+\lambda+\mu)} .$$

Using Appendix C, we obtain that

$$P_0(t) = \frac{\mu}{\lambda+\mu} + \frac{\lambda}{\lambda+\mu}e^{-(\lambda+\mu)t}.$$

Then

$$P_1(t) = \frac{\lambda}{\lambda + \mu} - \frac{\lambda}{\lambda + \mu} e^{-(\lambda+\mu)t}.$$

As t goes to infinity, $P_0(t)$ approaches $A = \mu/(\lambda + \mu)$. This is the stationary probability that the process is in state 0: $A = E[\tau_0]/(E[\tau_0] + E[\tau_1])$, where $\tau_0 \sim \text{Exp}(\lambda)$ and $\tau_1 \sim \text{Exp}(\mu)$.

Appendix A: Nonhomogeneous Poisson Process

A.1 Definition and Basic Properties

Definition A.1.1
The counting process $\{N(t),\ t \geq 0\}$ is said to be a *nonhomogeneous Poisson process* (NHPP) with intensity function $\lambda(t),\ t \geq 0$, if
(i) $N(0) = 0$;
(ii) $\{N(t),\ t \geq 0\}$ has independent increments;
(iii)$P(N(t+h) - N(t) \geq 2) = o(h)$ as $h \to 0$;
(iv) $P(N(t+h) - N(t) = 1) = \lambda(t)h + o(h)$ as $h \to 0$.

Denote by $\Lambda(t)$ the integral of the intensity function:

$$\Lambda(t) = \int_0^t \lambda(v)dv. \tag{A.1.1}$$

$\Lambda(t)$ is called the *cumulative event rate* or the *mean value function* of the process.

Theorem A.1.1

$$P(N(t+s) - N(t) = k) = e^{-\Lambda(t,s)} \frac{[\Lambda(t,s)]^k}{k!}, \quad k \geq 0, \tag{A.1.2}$$

where $\Lambda(t,s) = \Lambda(t+s) - \Lambda(t) = \int_t^{t+s} \lambda(v)dv$.

Proof.
Our goal is to find the functions

$$P_k(t,s) = P(N(t+s) - N(t) = k). \tag{A.1.3}$$

First consider $P_0(t, s+h)$. To have no events in $(t, t+s+h)$ means having no events in $(t, t+s)$ and no events in $(t+s, t+s+h)$. Due to (ii)–(iv),

$$P_0(t, s+h) = P_0(t,s)P_0(t+s,h) = P_0(t,s)[1-\lambda(t+s)h] + o(h). \quad \text{(A.1.4)}$$

Then

$$P_0(t, s+h) - P_0(t,s) = -P_0(t,s)\lambda(t+s)h + o(h).$$

Dividing both sides by h and letting $h \to 0$ yields

$$P_0'(t,s) = -P_0(t,s)\lambda(t+s) \; . \quad \text{(A.1.5)}$$

This differential equation must be solved for the initial condition $P_0(t,0) = 1$ since, with probability 1, there are no events in a zero-length interval. The solution is

$$P_0(t,s) = e^{-(\Lambda(t+s)-\Lambda(t))} = e^{-\int_t^{t+s}\lambda(v)dv}. \quad \text{(A.1.6)}$$

It is easy to show that for $k > 0$,

$$P_k(t, s+h) = P_k(t,s)P_0(t+s,h) + P_{k-1}(t,s)P_1(t+s,h) + o(h),$$

where $P_0(t+s,h) = 1 - \lambda(t+s)h + \mathrm{o}(h)$ and $P_1(t+s,h) = \lambda(t+s)h + o(h)$. Therefore,

$$\begin{aligned} P_k(t, s+h) = \; & P_k(t,s)[1-\lambda(t+s)h] \\ & + P_{k-1}(t,s)\lambda(t+s)h + o(h), \end{aligned} \quad \text{(A.1.7)}$$

from which it follows that

$$\frac{P_k(t,s+h) - P_k(t,s)}{h} = \lambda(t+s)[P_{k-1}(t,s) - P_k(t,s)] + o(1).$$

In the limit, for any $k > 0$

$$P_k'(t,s) = \lambda(t+s)[P_{k-1}(t,s) - P_k(t,s)]. \quad \text{(A.1.8)}$$

This system must be solved with the initial condition $P_k(t,0) = 0$ for $k > 0$.

It is instructive to demonstrate an efficient way of obtaining the solution. Introduce the following *generating function*:

$$F(t,s;x) = \sum_{k=0}^{\infty} P_k(t,s)x^k. \quad \text{(A.1.9)}$$

Multiply equation (A.1.8) by x^k and sum from $k = 0$ to infinity. (Set $P_{-1}(t,s) \equiv 0$.) Then, after some algebra, we obtain

$$\frac{\partial F}{\partial s} = (x-1)\lambda(t+s)F, \quad \text{(A.1.10)}$$

or

$$\frac{\partial \log F}{\partial s} = (x-1)\lambda(t+s). \tag{A.1.11}$$

It follows now from (A.1.11) that

$$\begin{aligned}\log F(t,s;x) - \log F(t,0;x) &= (x-1)\int_0^s \lambda(t+v)dv \\ &= (x-1)\int_t^{t+s} \lambda(y)dy \\ &= (x-1)\Lambda(t,s).\end{aligned} \tag{A.1.12}$$

Since $F(t,0;x) = P_0(t,0) = 1$, the solution of (A.1.12) is

$$F(t,s;x) = \exp[(x-1)\Lambda(t,s)] = \exp[-\Lambda(t,s)]\sum_{k=0}^{\infty}\frac{[\Lambda(t,s)]^k x^k}{k!}. \tag{A.1.13}$$

Comparing this expression with the definition (A.1.9) of $F(t,s;x)$, we see that

$$P_k(t,s) = \exp[-\Lambda(t,s)]\frac{[\Lambda(t,s)]^k}{k!},\ k \geq 0. \tag{A.1.14}$$

We see therefore that the number of events in $(t, t+s)$ in an NHPP follows a Poisson distribution with parameter $\Lambda(t,s)$. For a particular case of $\lambda(t) = \lambda$, we arrive at the result (2.1.6) for a Poisson process with rate λ.

Let us consider the distribution of the random intervals between two adjacent events in an NHPP. The situation for a Poisson process with constant rate λ was very simple: the interval between any two adjacent events is $\tau \sim \text{Exp}(\lambda)$. For an NHPP, the distribution of the distance between the event which took place at t^* and the next event depends on t^*.

Suppose that an event took place in an NHPP at the instant t. Denote by τ_t the interval to the next event.

Corollary A.1.1
(i) In an NHPP with intensity function $\lambda(t)$, the c.d.f. of τ_t is

$$P(\tau_t \leq x) = F_t(x) = 1 - \exp\Big[-\int_t^{t+x}\lambda(v)dv\Big], \tag{A.1.15}$$

and the density function of τ_t is

$$f_t(x) = \lambda(t+x)\exp\Big[-\int_t^{t+x}\lambda(v)dv\Big]. \tag{A.1.16}$$

(ii) The random variable τ_t does not depend on the history of the NHPP on the interval $[0,t]$.

Proof.
(i) Consider the interval $(t, t+x)$. By (A.1.6), $P_0(t,x) = e^{-\int_t^{t+x} \lambda(v)dv}$. If there are no events in $(t, t+x)$, then the interval to the next event exceeds x. This proves (A.1.15) because $F_t(x)$ is the probability of the complementary event. The density function $f_t(x)$ is a derivative of $F_t(x)$ with respect to x.
(ii) Follows from the assumptions (i)–(iv) of the NHPP.

For $\lambda(t) \equiv \lambda$, we arrive at the exponential distribution of the interval between adjacent events.

Note that the mean number of events on $[0,t]$ in an NHPP equals

$$\mu_{[o,t]} = \Lambda(t). \tag{A.1.17}$$

This follows directly from the fact that the number of events on $[0,t]$ has a Poisson distribution (A.1.14).

A.2 Parametric estimation of the intensity function

In this section we consider two important types of the intensity function – the so-called log-linear form and the Weibull form.

Log-linear form of the intensity function

Suppose that the intensity function of the NHPP has the form

$$\lambda(t) = e^{\alpha+\beta t}, \tag{A.2.1}$$

where α and β are unknown parameters. Following Cox and Lewis (1966), we describe the maximum likelihood approach to the estimation of α and β.

Suppose we observe an NHPP on the interval $[0,T]$, and suppose that the events took place at the instants $0 < t_1 < t_2 < \ldots < t_n < T$. Then by Corollary A.1.1, the likelihood function has the form:

$$\begin{aligned} Lik(t_1,\ldots,t_n) = \quad & f_0(t_1) f_{t_1}(t_2 - t_1) \ldots f_{t_{n-1}}(t_n - t_{n-1}) \\ & \times [1 - F_{t_n}(T - t_n)] \,. \end{aligned} \tag{A.2.2}$$

For $f_t(x)$ and $F_t(x)$ defined in Corollary A.1.1, the likelihood function takes the form

$$Lik(t_1,\ldots,t_n) = \prod_{i=1}^{n} \lambda(t_i) \exp\big[-\int_0^T \lambda(v)dv\big]. \tag{A.2.3}$$

Substitute $\lambda(t)$ from (A.2.1) into (A.2.2) and take the logarithm. Then, after simple algebra,

$$logLik(t_1, \ldots, t_n) = n\alpha + \beta \sum_{i=1}^{n} t_i - e^{\alpha}[e^{\beta T} - 1]/\beta \; . \tag{A.2.4}$$

The MLE of α and β will be found as the solution of the maximum likelihood equations

$$\partial logLik/\partial\alpha = n - e^{\alpha}(e^{\beta T} - 1)\beta^{-1} = 0 \tag{A.2.5}$$

and

$$\partial logLik/\partial\beta = \sum_{i=1}^{n} t_i - e^{\alpha}\Big(- (e^{\beta T} - 1)\beta^{-2} + e^{\beta} T \beta^{-1}\Big) = 0. \tag{A.2.6}$$

From (A.2.5) it follows that

$$e^{\hat{\alpha}} = \frac{n\hat{\beta}}{e^{\hat{\beta}T} - 1}. \tag{A.2.7}$$

Substituting $e^{\hat{\alpha}}$ from (A.2.7) into (A.2.6), we obtain, after some algebra, the equation for the MLE of β:

$$\sum_{i=1}^{n} t_i + n\hat{\beta}^{-1} = \frac{nT}{1 - e^{-\hat{\beta}T}}. \tag{A.2.8}$$

Suppose that we have observed *several* independent realizations of the NHPP with the same intensity function (A.2.1). The data from the ith realization are as follows: the events took place at the instants $t_1^{(i)}, \ldots, t_{n_i}^{(i)}$; the process was observed on the interval $[0, T_i]$, $i = 1, 2, \ldots, m$.

The "overall" likelihood function will be the product of m likelihood functions for separate realizations. The derivation of the equations for finding the MLEs is similar. We present the final result. The equation for $\hat{\beta}$ is

$$T_{tot} + \sum_{i=1}^{m} n_i/\hat{\beta} = \frac{\sum_{i=1}^{m} n_i \times \sum_{i=1}^{m} e^{\hat{\beta}T_i} T_i}{\sum_{i=1}^{m} e^{\hat{\beta}T_i} - m}, \tag{A.2.9}$$

where

$$T_{tot} = \sum_{j=1}^{n_1} t_j^{(1)} + \ldots + \sum_{j=1}^{n_m} t_j^{(m)}. \tag{A.2.10}$$

The equation for $\hat{\alpha}$ takes the form

$$e^{\hat{\alpha}} = \frac{\hat{\beta} \sum_{i=1}^{m} n_i}{\sum_{i=1}^{m} e^{\hat{\beta}T_i} - m}. \tag{A.2.11}$$

It is desirable to have a measure of the variability of $\hat{\alpha}$ and $\hat{\beta}$. We will derive the large-sample maximum likelihood confidence intervals described in Sect. 3.3. First we compute the second-order derivatives (multiplied by -1) evaluated at the MLEs $\hat{\alpha}$ and $\hat{\beta}$. They are

$$V_{11}(\hat{\alpha},\hat{\beta}) = n, \tag{A.2.12}$$

$$V_{12}(\hat{\alpha},\hat{\beta}) = \sum_{i=1}^{n} t_i \tag{A.2.13}$$

and

$$V_{22}(\hat{\alpha},\hat{\beta}) = \hat{\beta}^{-1}\Big(\sum_{i=1}^{n} t_i(\hat{\beta}T-2)+nT\Big). \tag{A.2.14}$$

V_{ij} are the elements of the observed Fisher information matrix $\mathbf{I}_F$. The elements of its inverse $\mathbf{I}_F^{-1}$ are:

$$\hat{W}_{11} = V_{22}/\big(V_{22}n - \big(\sum_{i=1}^{n} t_i\big)^2\big); \tag{A.2.15}$$

$$\hat{W}_{12} = -(V_{12}/V_{22})\hat{W}_{11}\ ; \tag{A.2.16}$$

$$\hat{W}_{22} = n/\big(V_{22}n - \big(\sum_{i=1}^{n} t_i\big)^2\big). \tag{A.2.17}$$

Therefore, the estimates of standard errors of $\hat{\alpha}$ and $\hat{\beta}$ are

$$\hat{\sigma}_{\hat{\alpha}} = \sqrt{\hat{W}_{11}},\ \hat{\sigma}_{\hat{\beta}} = \sqrt{\hat{W}_{22}}. \tag{A.2.18}$$

Weibull-form intensity function

Suppose that the intensity function has the form

$$\lambda(t) = \alpha^{\beta}\beta t^{\beta-1}. \tag{A.2.19}$$

For this intensity function, the time to the first failure $\tau_0 \sim W(\alpha,\beta)$, as follows directly from (A.1.15). The loglikelihood now is

$$logLik(t_1,\ldots,t_n) = n\beta\log\alpha + n\log\beta + (\beta-1)\sum_{i=1}^{n}\log t_i - (\alpha T)^{\beta}\ . \tag{A.2.20}$$

The MLE will be found as the solution of the following two equations:

$$\partial logLik/\partial\alpha = n\beta/\alpha - \beta\alpha^{\beta-1}T^{\beta} = 0 \tag{A.2.21}$$

and

$$\partial logLik/\partial\beta = n\log\alpha + n/\beta + \sum_{i=1}^{n}\log t_i - (\alpha T)^{\beta}\beta\log(\alpha T) = 0. \quad \text{(A.2.22)}$$

The solution is:

$$\hat{\alpha} = n^{1/\hat{\beta}}/T, \;\; \hat{\beta} = \Big(\log T - n^{-1}\sum_{i=1}^{n}\log t_i)\Big)^{-1}. \quad \text{(A.2.23)}$$

The elements of the Fisher information matrix are:

$$V_{11}(\hat{\alpha},\hat{\beta}) = n\hat{\beta}^2/\hat{\alpha}^2, \quad \text{(A.2.24)}$$

$$V_{12}(\hat{\alpha},\hat{\beta}) = \hat{\beta}n\log(\hat{\alpha}T)/\hat{\alpha} \quad \text{(A.2.25)}$$

and

$$V_{22}(\hat{\alpha},\hat{\beta}) = n/\hat{\beta}^2 + n(\log(\hat{\alpha}T))^2. \quad \text{(A.2.26)}$$

From here it follows that the estimates of standard errors of $\hat{\alpha}$ and $\hat{\beta}$ are

$$\hat{\sigma}_{\hat{\alpha}} = \sqrt{V_{22}/(V_{11}V_{22} - V_{12}^2)}, \;\; \hat{\sigma}_{\hat{\beta}} = \sqrt{V_{11}/(V_{11}V_{22} - V_{12}^2)}. \quad \text{(A.2.27)}$$

Remark
Suppose that the point process $N(t)$, $t \geq 0$, describes failures of a renewable system. In this context, the derivative of the mean number of failures on $[0,t]$ with respect to t,

$$v(t) = \frac{E[N(t)]}{dt}, \quad \text{(A.2.28)}$$

is called *the rate of occurrence of failures* (ROCOF). This term is very popular in reliability literature; see e.g. O'Connor (1991), Crowder et al (1991). Obviously, for an NHPP, the ROCOF equals the intensity function $\lambda(t)$; see (A.1.16).

The ROCOF is often confused with the *failure rate* $h(t)$ defined earlier in Chap. 2. The principal difference is that the notion of failure rate is defined only for a random variable describing the lifetime of a nonrepairable component (system). In an NHPP, the failure rate of the time to the first failure τ_0 *coincides* with $\lambda(t)$ only on the interval $[0,\tau_0]$.

More information on statistical inference for the NHPP process, including simple graphical analysis, can be found Cox and Lewis (1966) and Crowder et al (1991).

Appendix B: Covariances and Means of Order Statistics

Table B.1(a)

Covariance matrix of order statistics for $n = 8$, $Z \sim \text{Extr}(0, 1)$

i	1	2	3	4	5	6	7	8
1	1.645	0.422	0.262	0.180	0.131	0.097	0.071	0.048
2		0.464	0.280	0.193	0.140	0.103	0.076	0.052
3			0.398	0.201	0.152	0.112	0.082	0.056
4				0.290	0.168	0.124	0.090	0.062
5					0.232	0.140	0.102	0.070
6						0.198	0.121	0.083
7							0.183	0.106
8								0.200

Table B.1(b)

Mean values of the order statistics

$m_{(1)}$	$m_{(2)}$	$m_{(3)}$	$m_{(4)}$	$m_{(5)}$	$m_{(6)}$	$m_{(7)}$	$m_{(8)}$
−2.6567	−1.5884	−1.0111	−0.5880	−0.2312	0.1029	0.4548	0.9021

Table B.2(a).

The covariance matrix of order statistics for $n = 10, r = 9,\ Z \sim \mathrm{Extr}(0,1)$.

i	1	2	3	4	5	6	7	8	9
1	1.645	0.436	0.275	0.193	0.144	0.111	0.086	0.067	0.051
2		0.646	0.290	0.204	0.152	0.117	0.091	0.071	0.054
3			0.397	0.217	0.162	0.124	0.097	0.076	0.058
4				0.287	0.174	0.137	0.104	0.081	0.062
5					0.227	0.145	0.113	0.088	0.067
6						0.190	0.125	0.098	0.074
7							0.166	0.111	0.085
8								0.152	0.100
9									0.149

Table B.2(b)

Mean values of the order statistics, $n = 10$

$m_{(1)}$	$m_{(2)}$	$m_{(3)}$	$m_{(4)}$	$m_{(5)}$
−2.800	−1.826	−1.267	−0.868	−0.544
$m_{(6)}$	$m_{(7)}$	$m_{(8)}$	$m_{(9)}$	$m_{(10)}$
−0.257	−0.012	0.284	0.585	0.990

Table B.3(a)

The covariance matrix of order statistics for $n = 15,\ r = 10,\ Z \sim \mathrm{Extr}(0,1)$.

i	1	2	3	4	5	6	7	8	9	10
1	1.645	0.455	0.293	0.211	0.162	0.129	0.106	0.088	0.074	0.062
2		0.645	0.303	0.219	0.168	0.134	0.109	0.091	0.076	0.064
3			0.396	0.227	0.174	0.139	0.114	0.094	0.079	0.067
4				0.285	0.182	0.145	0.118	0.098	0.082	0.069
5					0.223	0.152	0.124	0.103	0.086	0.073
6						0.184	0.130	0.108	0.091	0.077
7							0.157	0.115	0.096	0.081
8								0.138	0.103	0.087
9									0.124	0.093
10										0.113

Table B.3(b)

Mean values of the order statistics

$m_{(1)}$	$m_{(2)}$	$m_{(3)}$	$m_{(4)}$	$m_{(5)}$
−3.285	−2.250	−1.713	−1.340	−1.048

$m_{(6)}$	$m_{(7)}$	$m_{(8)}$	$m_{(9)}$	$m_{(10)}$
−0.802	−0.585	−0.387	−0.201	−0.021

Table B.4(a)

The covariance matrix of order statistics for $n = 8,\ Z \sim N(0,1)$.

i	1	2	3	4	5	6	7	8
1	0.373	0.186	0.126	0.095	0.075	0.060	0.048	0.037
2		0.239	0.163	0.123	0.098	0.079	0.063	0.048
3			0.200	0.152	0.121	0.098	0.079	0.060
4				0.187	0.149	0.121	0.098	0.075
5					0.187	0.152	0.123	0.095
6						0.200	0.163	0.126
7							0.239	0.186
8								0.373

Table B.4(b)

Mean values of the order statistics

$m_{(1)}$	$m_{(2)}$	$m_{(3)}$	$m_{(4)}$
−1.4236	−0.85224	−0.4728	−0.1525

$m_{(4+k)} = -m_{(4-k+1)},\ k = 1,2,3,4$

Table B.5(a)

The covariance matrix of order statistics for $n = 10,\ Z \sim N(0,1)$.

i	1	2	3	4	5	6	7	8	9	10
1	0.344	0.171	0.116	0.088	0.071	0.058	0.049	0.041	0.034	0.027
2		0.214	0.147	0.112	0.090	0.074	0.062	0.052	0.043	0.034
3			0.175	0.134	0.108	0.089	0.075	0.063	0.052	0.041
4				0.158	0.128	0.106	0.089	0.075	0.062	0.049
5					0.151	0.126	0.106	0.089	0.074	0.058
6						0.151	0.128	0.108	0.090	0.071
7							0.158	0.138	0.112	0.088
8								0.175	0.147	0.116
9									0.214	0.171
10										0.344

Table B.5(b)

Mean values of the order statistics

$m_{(1)}$	$m_{(2)}$	$m_{(3)}$	$m_{(4)}$	$m_{(5)}$
−1.5388	−1.0014	−0.6561	−0.3758	−0.1227

$m_{(5+k)} = -m_{(5-k+1)},\ k = 1, 2, 3, 4, 5$

Table B.6(a)

The covariance matrix of order statistics for $n = 15,\ r = 10,\ Z \sim N(0, 1)$.

i	1	2	3	4	5	6	7	8	9	10
1	0.301	0.148	0.101	0.077	0.063	0.053	0.45	0.040	0.035	0.031
2		0.179	0.122	0.094	0.076	0.064	0.055	0.048	0.043	0.038
3			0.141	0.108	0.088	0.074	0.064	0.056	0.049	0.044
4				0.122	0.100	0.084	0.073	0.064	0.056	0.050
5					0.112	0.095	0.082	0.071	0.063	0.056
6						0.106	0.091	0.080	0.071	0.063
7							0.103	0.090	0.080	0.071
8								0.102	0.090	0.080
9									0.103	0.091
10										0.106

Table B.6(b)

Mean values of the order statistics

$m_{(1)}$	$m_{(2)}$	$m_{(3)}$	$m_{(4)}$	$m_{(5)}$	$m_{(6)}$	$m_{(7)}$	$m_{(8)}$
−1.736	−1.248	−0.948	−0.715	−0.516	−0.335	−0.165	−0.000

For $m_{(8+k)} = -m_{(8-k)},\ k = 1, 2, 3, 4, 5, 8, 7$

Appendix C: The Laplace Transform

The Laplace transform[1] $\pi(s)$ of a nonnegative function $g(t)$ is defined as

$$\pi(s) = \int_0^\infty e^{-st} g(t) dt \, .$$

$\pi(s)$	$g(t)$	
$\frac{1}{s+\lambda}$	$e^{-\lambda s}$	$(C.1)$
μ	μ/s	$(C.2)$
$\frac{1}{1+as}$	$e^{-t/a}/a$	$(C.3)$
$\frac{1}{s(s-a)}$	$(e^{at}-1)/a$	$(C.4)$
$\frac{1}{(s-a)^2}$	te^{ta}	$(C.5)$
$\frac{1}{(s-a)(s-b)}, \ a \neq b$	$\frac{e^{as}-e^{bs}}{a-b}$	$(C.6)$
$\frac{s}{(s-a)^2}$	$(1+as)e^{as}$	$(C.7)$
$\frac{s}{(s-a)(s-b)}$	$\frac{ae^{as}-be^{bs}}{a-b}$	$(C.8)$
$\left(\frac{\lambda}{\lambda+s}\right)^k$	$\frac{\lambda^k s^{k-1} e^{-\lambda s}}{(k-1)!}$	$(C.9)$
$\frac{1}{(s-a)(s-b)(s-c)}$	$\frac{(c-b)e^{sa}+(a-c)e^{sb}+(b-a)e^{sc}}{(a-b)(a-c)(c-b)}$	$(C.10)$

[1] reproduced from Gertsbakh (1989, p. 281)by courtesy of Marcel Dekker Inc

Appendix D: Probability Paper

D.1. NORMAL PROBABILITY PAPER

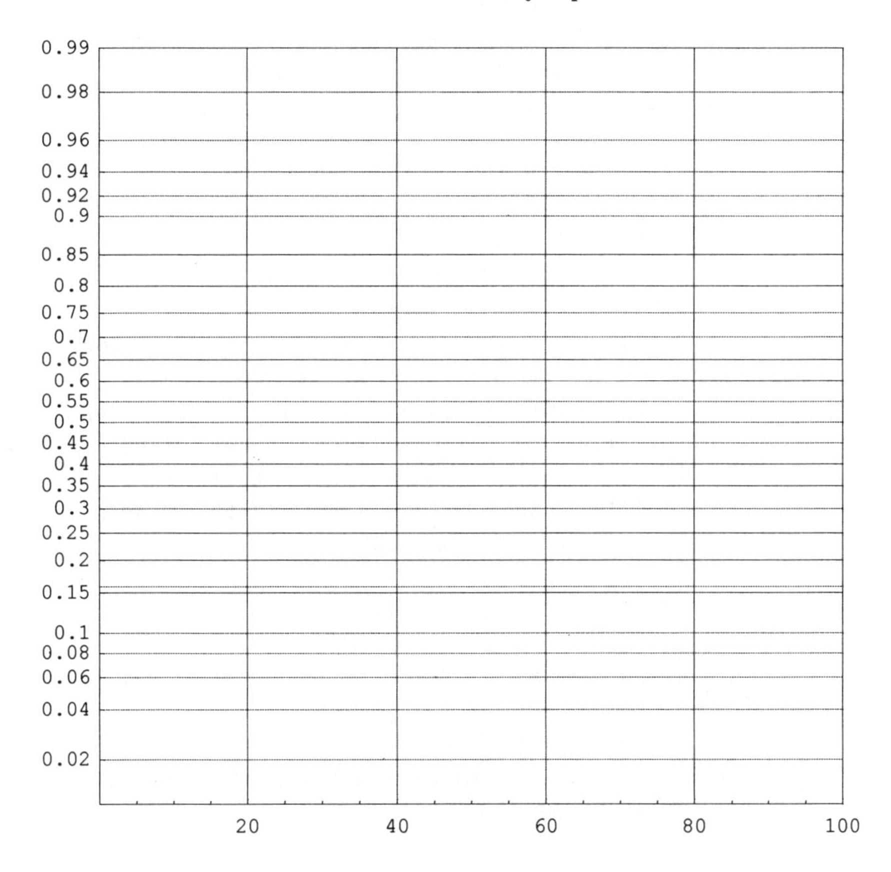

Below is the *Mathematica* code for producing normal paper and for plotting on it. Data processing using normal paper is described in Example 3.2.2.

```
In[214]:= Needs["Statistics`ContinuousDistributions`"]
          Needs["Graphics`MultipleListPlot`"]
          dist = NormalDistribution[0, 1];
          y = Quantile[dist, 0.99] + 2.326;
          tn = {3, 14, 25, 38, 40, 51, 66};
          (*complete observations*)
          nobsn = 10; (*the total number of pobservations*)
          (*++++++++++++++++++++++++++++++++++++++++++++*)
          tn = Sort[tn];
          r = Length[tn];
          For[i = 1, i < r + 1, i++, d1[i] = tn[[i]];
            d2[i] = 2.326 + Quantile[dist, (i - 0.5) / nobsn]];
          xydatan = Table[{d1[i], d2[i]}, {i, 1, r}];
          pict1 = ListPlot[xydatan,
            AspectRatio → 1, PlotStyle → PointSize[0.02],
            PlotRange → {{0, 100}, {0, 4.652}},
            GridLines → {Automatic,
              {0.00, 0.272, 0.575, 0.771, 0.921, 1.044,
               1.2896, 1.326, 1.486, 1.651, 1.8016,
               1.941, 2.073, 2.2, 2.326,
               2.452, 2.579, 2.711, 2.850, 3.00, 3.168, 3.362,
               3.608, 3.731, 3.881, 4.077, 4.380, 4.652}},
            Ticks → {Automatic, {{0.272, 0.02}, {0.575, 0.04},
               {0.771, 0.06}, {0.921, 0.08}, {1.044, 0.10},
               {1.2896, 0.15}, {1.484, 0.20}, {1.651, 0.25},
               {1.8016, 0.30}, {1.941, 0.35}, {2.073, 0.40},
               {2.2, 0.45}, {2.326, 0.50},
               {2.452, 0.55}, {2.579, 0.60},
               {2.711, 0.65}, {2.85, 0.70},
               {3.00, 0.75}, {3.168, 0.80},
               {3.362, 0.85}, {3.608, 0.90}, {3.731, 0.92},
               {3.881, 0.94}, {4.077, 0.96},
               {4.380, 0.98}, {4.652, 0.99}}},
            PlotLabel → FontForm["Normal Probability Paper \n",
              {"Times-Bold", 12}]]
```

D.2. WEIBULL PROBABILITY PAPER

Weibull Probability Paper

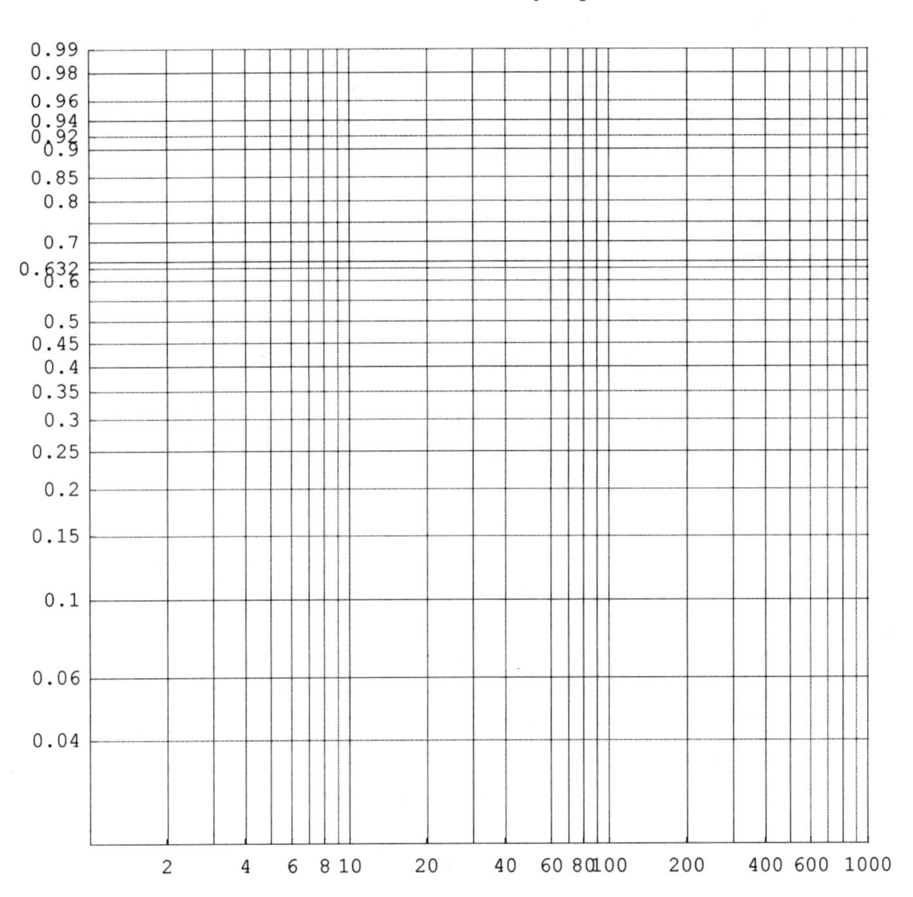

Below is the *Mathematica* code for producing Weibull paper and for plotting on it. Data processing using Weibull paper is described in Example 3.2.1.

```
Needs["Graphics`MultipleListPlot`"]
tw = {2.25, 6.7, 37.6, 85.4, 110};
(*complete observations*)
nobsw = 7; (*the total number of pobservations*)
(*+++++++++++++++++++++++++++++++++++++++++++*)
a = 0; c = 2.302; d = 4.605; e = 3.9;
tw = Sort[tw]; r = Length[tw];
For[i = 1, i < r + 1, i++, d1w[i] = a + Log[tw[[i]]];
  d2w[i] = 3.9 + Log[-Log[1 - (i - 0.5) / nobsw]]];
xydataw = Table[{d1w[i], d2w[i]}, {i, 1, r}];
pict1 = ListPlot[xydataw, AspectRatio → 1,
   PlotStyle → PointSize[0.02],
   PlotRange → {{0, Log[1000] + 0.002}, {-3.9 + e, 1.527 + e}},
   GridLines -> {{0, 0.693, 1.099, 1.386, 1.609, 1.791, 1.946,
      2.079, 2.2, c, 0.693 + c, 1.099 + c, 1.386 + c,
      1.791 + c, 1.946 + c, 2.079 + c, 2.2 + c, d, 0.693 + d,
      1.099 + d, 1.386 + d, 1.609 + d, 1.791 + d, 1.946 + d,
      2.079 + d, 2.2 + d, 2.302 + d},
     {0, -3.2 + 3.9, -2.78 + 3.9, -2.25 + e, -1.82 + e,
      -1.5 + e, -1.245 + e, -1.031 + e, -0.842 + e,
      -0.672 + e, -0.514 + e, -0.366 + e, -0.225 + e,
      -0.087 + e, 0.00 + e, 0.0486 + e, 0.186 + e,
      0.327 + 3.90, 0.476 + e, 0.640 + e, 0.834 + e, 0.926 + e,
      1.034 + e, 1.169 + e, 1.364 + 3.9, 1.527 + e}},
   Ticks -> {{{0.683, 2}, {1.38, 4}, {1.791, 6}, {2.079, 8},
      {2.302, 10}, {0.683 + 2.3, 20}, {1.38 + 2.3, 40},
      {1.791 + 2.3, 60}, {2.079 + 2.3, 80}, {4.605, 100},
      {0.683 + 4.605, 200}, {1.38 + 4.605, 400},
      {1.791 + 4.605, 600}, {6.908, 1000}},
     {{0, 0.02}, {0.7, 0.04}, {-2.78 + e, 0.06},
      {-2.25 + e, 0.10}, {-1.817 + e, 0.15}, {-1.500 + e, 0.20},
      {-1.246 + e, 0.25}, {-1.031 + e, 0.3}, {-0.842 + e, 0.35},
      {-0.672 + 3.9, 0.4}, {-0.514 + e, 0.45}, {-0.366 + e, 0.50},
      {-0.087 + e, 0.60}, {0.00 + e, 0.632}, {0.186 + e, 0.70},
      {0.476 + e, 0.80}, {0.64 + e, 0.85}, {0.834 + e, 0.90},
      {0.926 + e, 0.92}, {1.034 + e, 0.94}, {1.169 + e, 0.96},
      {1.364 + e, 0.98}, {1.527 + e, 0.99}}},
  PlotLabel -> FontForm["Weibull Probability Plot \n",
    {"Times-Bold", 12}]] (*To get the Weibull
   paper without ploted points, put a=10*)
```

Appendix E: Renewal Function

Table E.1

The renewal function $m(t)$ for $F(t) = 1 - e^{-t^{\beta}}$

t	$\beta = 1.5$	$\beta = 2$	$\beta = 2.5$	$\beta = 3$	$\beta = 3.5$	$\beta = 4$
0.05	0.011	0.003	0.001	0.000	0.000	0.000
0.10	0.032	0.010	0.003	0.001	0.000	0.000
0.15	0.060	0.023	0.009	0.004	0.001	0.001
0.20	0.093	0.042	0.019	0.008	0.004	0.002
0.25	0.130	0.066	0.033	0.017	0.008	0.004
0.30	0.172	0.096	0.053	0.029	0.016	0.009
0.35	0.216	0.130	0.078	0.046	0.027	0.016
0.40	0.264	0.170	0.109	0.069	0.044	0.028
0.45	0.314	0.215	0.146	0.098	0.066	0.044
0.50	0.367	0.265	0.189	0.134	0.095	0.067
0.55	0.422	0.318	0.238	0.178	0.132	0.098
0.60	0.477	0.376	0.294	0.229	0.178	0.138
0.65	0.537	0.437	0.355	0.288	0.233	0.189
0.70	0.597	0.502	0.421	0.354	0.297	0.249
0.75	0.657	0.569	0.492	0.427	0.370	0.321
0.80	0.719	0.639	0.568	0.509	0.452	0.404
0.85	0.782	0.711	0.647	0.591	0.541	0.497
0.90	0.845	0.785	0.729	0.680	0.637	0.599
0.95	0.909	0.860	0.814	0.773	0.738	0.707
1.00	0.973	0.936	0.890	0.868	0.842	0.820

References

Abernethy, R.B., Breneman, J.E., Medlin, C.H. and G.L. Reinman. 1983. *Weibull Analysis Handbook.* Air Force Wright Aeronautical Laboratories Technical Report AFWAL-TR-83-2079.

Aitchison, J. and J.A.C. Brown. 1957. *The Lognormal Distribution.* Cambridge University Press, New York and London.

Anderson, T.W. 1984. *An Introduction to Multivariate Statistical Analysis*, 2nd ed. New York: Wiley.

Andronov, A.M. 1994. Analysis of nonstationary inifinite-linear qeueing system. *Automat. Control Comput. Sci.*, **28**, 28-33.

Andronov, A.M. and I. Gertsbakh. 1972. Optimum maintenance in a certain model of accumulation of damages. *Engrg. Cybern.*, **10**(5), 620-628.

Artamanovsky, A.V. and Kh.B. Kordonsky. 1970. Estimate of maximum likelihood for simplest grouped data. *Theory Probab. Appl.*, **15**, 128-132.

Arunkumar, S.A. 1972. Nonparametric age replacement. *Sankhyā* **A**, **34**, 251-256.

Bailey, R. and B. Mahon. 1975. A proposed improved replacement policy for army vehicles. *Oper. Res. Quart.* **26**, 477-494.

Barlow, R.E. 1998. *Engineering Reliability.* Philadelphia: Society of Industrial and Applied Mathematics.

Barlow, R.E. and A. Marshall. 1964. Bounds for distributions with monotone hazard rate, I and II. *Ann. Math. Statist.*, **35**, 1234-1274.

Barlow, R.E. and F. Proschan. 1975. *Statistical Theory of Reliability and Life Testing.* New York: Holt, Rinehart and Winston.

Bellman, R.E. 1957. *Dynamic Programming.* Princeton, NJ: Princeton University Press.

Burtin, Yu. and B. Pittel. 1972. Asymptotic estimates of the reliability of a complex system. *Engrg. Cybern.*, **10**(3), 445-451.

Cox, D.R. and P.A.W. Lewis. 1966. *The Statistical Analysis of Series of Events.* London: Chapman and Hall.

Crowder, M.J., Kimber, A.C., Smith, R.L. and T.J. Sweeting. 1991. *Statistical Analysis of Reliability Data.* New York: Chapman and Hall.

DeGroot, M. H. 1970. *Optimal Statistical Decisions.* McGraw-Hill Company, New York.

Devore, J. L. 1982. *Probability and Statistics for Engineering and the Sciences.* Brooks/Cole Publishing Company, Monterey, California.

Dodson, B. 1994. *Weibull Analysis.* ASQC Quality Press, Milwaukee, Wisconsin.

Elperin, T. and I. Gertsbakh. 1987. Maximum likelihood estimation in a Weibull regression model with type I censoring: A Monte Carlo study. *Comm. Statist., Simulation Computat.*, **18**, 349-372.

Elperin, T., I. Gertsbakh and M. Lomonosov. 1991. Estimation of network reliability using graph evolution models. *IEEE Trans. Reliab.*, **40**(5), 572-581.

Elsayed, E.A. 1996. *Reliability Engineering.* Addison Wesley Longman, Inc. Reading, MA.

Feller, W. 1968.*Introduction to Probability Theory and Its Applications*, Vol. 1, 3rd ed. New York: Wiley.

Fishman, G.S. 1996. *Monte Carlo: Concepts, Algorithms and Applications.* New York: Springer.

George, L., H. Mahlooji and Po-Wen Hu. 1979. Optimal replacement and build policies. In *Proceedings 1979 Ann. Reliability and Maintainability Symposium.* New York: Institute of Electrical and Electronic Engineers.

Gertsbakh, I. 1972. Preventive maintenance of objects with multi-dimensional state description. *Engrg. Cybern.*, **10** No. 5, 91-95 (in Russian).

Gertsbakh, I. 1977. *Models of Preventive Maintenance.* North-Holland, Amsterdam–New York–Oxford.

Gertsbakh, I. 1984. Optimal group preventive maintenance of a system with observable state parameter. *J. Appl. Prob.*, **16**, 923-925.

Gertsbakh, I. 1989. . *Statistical Reliability Theory.* New York: Marcel Dekker.

Gertsbakh, I. and Kh. Kordonsky. 1969. *Models of Failure.* Berlin Heidelberg New York: Springer.

Gertsbakh, I. and Kh. Kordonsky. 1994. The best time scale for age replacement. *Internat. J. Reliab., Quality Safety Engrg.* **1**, 219-229.

Gnedenko, B.V., Yu.K. Belyaev and A.D. Solovyev. 1969. *Mathematical Methods in Reliability Theory.* New York: Academic Press.

Hastings, C. 1969. The repair limit replacement method. *Oper. Res. Quart.*, **20**, 337-350.

Johnson, R.A. and D.W. Wichern. 1988. *Applied Multivariate Statistical Analysis*, 2nd ed. Prentice Hall, Englewood Cliffs, New Jersey 07632.

Kaplan, E.L. and P. Meier. 1958. Nonparametric estimation from incomplete observations. *J. Amer. Statist. Assoc.*, **53**, 457-481.

Khinchine, A.Ya. 1956. Streams of random events without aftereffect. *Theory Probab. Appl.*, **1**, 1-15.

Kordonsky, Kh.B. 1966. Statistical analysis of fatigue experiments. Private communication, Riga.

Kordonsky, Kh.B. and I. Gertsbakh. 1993. Choice of the best time scale for system reliability analysis. *European J. Oper. Res.*, **65**, 235-246.

Kordonsky, Kh.B and I. Gertsbakh. 1997. Fatigue crack monitoring in parallel time scales. *Proceedings of ESREL*, Lisboa, pp. 1485-1490. Kluwer Academic Publishers, Dordrecht Boston London.

Kordonsky, Kh.B and I. Gertsbakh. 1998. Parallel time scales and two-dimensional manufacturer and individual customer warranties. *IEEE Transactions*, **30**, 1181-1189.

Kovalenko, I.N., N.Yu. Kuznetsov and Ph.A. Pegg. 1997. *Mathematical Theory of Reliability of Time-Dependent Systems.* New York: Wiley.

Kullback, S. 1959. *Information Theory and Statistics.* New York: Wiley.

Kulldorf, G. 1961. *Estimation from Grouped and Partially Grouped Data.* New York: Wiley.

Lawless, J.F. 1982. *Statistical Models and Methods for Lifetime Data.* New

York: Wiley.

Lawless, J.F. 1983. Statistical methods in reliability. *Technometrics*, **25**, 305-316.

Lieblein, J. and M. Zelen. 1956. Statistical investigation of the fatigue life of deep groove ball bearings. *J. Res. Nat. Bur. Stand.*, **57**, 273-316.

Mann, N.R. and K.F. Fertig. 1973. Tables for obtaining confidence bounds and tolerance bounds based on best linear invariant estimates of parameters of the extreme value distribution. *Technometrics*, **17**, 87-101.

O'Connor, P.D.T. 1991. *Practical Reliability Engineering*, 3rd ed. New York: Wiley and Sons. .

Okumoto, E. and E.A. Elsayed. 1983. An optimum group maintenance policy. *Naval Res. Logist.*, **30**, 667-674.

Rappaport, A. 1998. *Decision Theory and Decision Behaviour.* McMillan Press.

Redmont, E.F., A.H. Christer, S.R. Ridgen, E. Burley, A. Tajelli and A. Abu-Tair. 1997. OR modeling of the deterioration and maintenance of concrete structures. *Eropean. J. Oper. Res.*, **99**, 619-631.

Ross, S.M. 1970. *Applied Probability Models with Optimization Applications.* San Francisco: Holden Day.

Ross, S.M. 1993. *Introduction to Probability Models.* 5th ed. New York: Academic Press.

Sarhan, A.E. and B.G. Greenberg. 1962. The best linear estimates for the parameters of the normal distribution. In *Contributions to Order Statistics*; A.E. Sarhan (ed.). New York: Wiley.

Seber, G.A.F. 1977. *Linear Regression Analysis.* New York: Wiley and Sons.

Stoikova, L.S. and I.N. Kovalenko. 1986. On certain extremal problems of reliability theory. *Tekhn. Kibernetica*, **6**, 19-23 (in Russian).

Taylor, H.M. and S. Karlin. 1984. *An Introduction to Stochastic Modeling.* New York: Academic Press.

Usher, J. S., A.H. Kamal, and W.H. Syed. 1998. Cost optimal preventive maintenance and replacement scheduling. *IIE Trans.*, **30**, 1121-1128.

Vardeman, S. B. 1994. *Statistics for Engineering Problem Solving.* Boston: PWS.

Williams, J.H., A. Davies and P.R. Drake. 1998. *Condition-Based Maintenance and Machine Diagnostics*. London: Chapman and Hall.

Zhang, F. and A. K.S. Jardine. 1998. Optimal maintenance models with minimal repair, periodic overhaul and complete renewal. *IIE Trans.*, **30**, 1109-1119.

Glossary of Notation

c.d.f.	Cumulative distribution function
d.f.	Density function
i.i.d.	Independent identically distributed
r.v.	Random variable
DFR	Decreasing failure rate
ER	Emergency repair
IFR	Increasing failure rate
NHPP	Nonhomogeneous Poisson Process
PM	Preventive maintenance
SMP	semi-Markov process

$X \sim F(x)$	X has c.d.f. $F(x)$.
$X \sim f(x)$	X has d.f. $f(x)$.
$\tau \sim B(n,p)$	τ has binomial distribution with parameters n and p.
$\tau \sim Geom(p)$	τ has geometric distribution with parameter p.
$\tau \sim \text{Exp}(\lambda)$	τ has exponential distribution with parameter λ.
$\tau \sim \text{Exp}(a,\lambda)$	τ has exponential distribution with parameters a, λ.
$\tau \sim \text{Extr}(a,b)$	τ has the extreme-value distribution with parameters a, b.
$\tau \sim Gamma(k,\lambda)$	τ has gamma distribution with parameters k, λ.
$\tau \sim logN(\mu,\sigma)$	τ has lognormal distribution with parameters μ, σ.
$\tau \sim N(\mu,\sigma)$	τ has normal distribution with parameters μ, σ.
$\tau \sim \mathcal{P}(\mu)$	τ has Poisson distribution with parameter μ.
$\tau \sim U(a,b)$	τ is uniformly distributed on $[a,b]$.
$\tau \sim W(\lambda,\beta)$	τ has Weibull distribution with parameters λ, β.
$E[X]$	The mean (expected) value of r.v. X.

$Var[X]$	The variance of r.v. X.
$c.v.[X]$	The coefficient of variation of X, i.e. $c.v.[X] = \sqrt{Var[X]}/E[X]$.
t_p	The p quantile, i.e. $P(X \le t_p) = p$.
$\lambda(t)$	Event rate in an NHPP: $\lambda(t)\Delta \approx P(\text{one event in } (t, t+\Delta))$.
$h(t)$	Failure (hazard) rate: $h(t) = f(t)/(1 - F(t))$.
$m(t)$	Renewal function.
$m'(t)$	Renewal density, i.e. $m'(t) = dm(t)/dt$.
$\log b$	Natural logarithm of b.
$\Lambda(t) = \int_0^t \lambda(v)dv$	Cumulative event rate in an NHPP.

$\mathbf{x} < \mathbf{y}$, where $\mathbf{x} = (x_1, x_2, \ldots, x_n)$, $\mathbf{y} = (y_1, \ldots, y_n)$, means that $x_i \le y_i$, $i = 1, \ldots, n$, and there is one coordinate j for which $x_j < y_j$.

Index

…ing: Mercedes-Druck, Berlin
… Buchbinderei Lüderitz & Bauer, Berlin